Advanced Ceramic Coatings and Interfaces V

Advanced Ceramic Coatings and Interfaces V

*A Collection of Papers Presented at the
34th International Conference on Advanced
Ceramics and Composites
January 24–29, 2010
Daytona Beach, Florida*

Edited by
Dongming Zhu
Hua-Tay Lin

Volume Editors
Sanjay Mathur
Tatsuki Ohji

The
American
Ceramic
Society

A John Wiley & Sons, Inc., Publication

Published by John Wiley & Sons, Inc., Hoboken, New Jersey.
Published simultaneously in Canada.

For general information on our other products and services or for technical support, please contact our Customer Care Department within the United States at (800) 762-2974, outside the United States at (317) 572-3993 or fax (317) 572-4002.

Wiley also publishes its books in a variety of electronic formats. Some content that appears in print may not be available in electronic format. For information about Wiley products, visit our web site at www.wiley.com.

Library of Congress Cataloging-in-Publication Data is available.

ISBN 978-0-470-59468-1

Printed in the United States of America.

10 9 8 7 6 5 4 3 2 1

Contents

v

ADVANCED COATINGS FOR WEAR AND CORROSION APPLICATIONS

ADVANCED COATING PROCESSING

Preface

The Symposium on Advanced Ceramic Coatings for Structural, Environmental and Functional Applications was held at the 34th Cocoa Beach International Conference on Advanced Ceramics and Composites in Cocoa Beach, Florida, during January 24–29, 2010. A total of 52 papers, including 15 invited talks, were presented at the symposium, covering broad ceramic coating and interface topic areas and emphasizing the latest advancement in coating processing, characterization and development.

The present volume contains fifteen contributed papers from the symposium, with topics including advanced coating processing, advanced coating for wear, corrosion, and oxidation resistance, and thermal and mechanical properties, highlighting the state-of-the-art ceramic coatings technologies for various critical engineering applications.

We are greatly indebted to the members of the symposium organizing committee, including Uwe Schulz, Yutaka Kagawa, Rodney Trice, Irene Spitsberg, Dileep Singh, Brain Hazel, Robert Vaßen, Sophoclis Patsias, Yong-Ho Sohn, Ping Xiao, and Jung Xu, for their assistance in developing and organizing this vibrant and cutting-edge symposium. We also would like to express our sincere thanks to manuscript authors and reviewers, all the symposium participants and session chairs for their contributions to a successful meeting. Finally, we are also grateful to the staff of The American Ceramic Society for their efforts in ensuring an enjoyable conference and the high-quality publication of the proceeding volume.

DONGMING ZHU, NASA Glenn Research Center, Cleveland, Ohio
H. T. LIN, Oak Ridge National Laboratory, Oak Ridge, Tennessee

Introduction

This CESP issue represents papers that were submitted and approved for the proceedings of the 34th International Conference on Advanced Ceramics and Composites (ICACC), held January 24–29, 2010 in Daytona Beach, Florida. ICACC is the most prominent international meeting in the area of advanced structural, functional, and nanoscopic ceramics, composites, and other emerging ceramic materials and technologies. This prestigious conference has been organized by The American Ceramic Society's (ACerS) Engineering Ceramics Division (ECD) since 1977.

The conference was organized into the following symposia and focused sessions:

Symposium 1	Mechanical Behavior and Performance of Ceramics and Composites
Symposium 2	Advanced Ceramic Coatings for Structural, Environmental, and Functional Applications
Symposium 3	7th International Symposium on Solid Oxide Fuel Cells (SOFC): Materials, Science, and Technology
Symposium 4	Armor Ceramics
Symposium 5	Next Generation Bioceramics
Symposium 6	International Symposium on Ceramics for Electric Energy Generation, Storage, and Distribution
Symposium 7	4th International Symposium on Nanostructured Materials and Nanocomposites: Development and Applications
Symposium 8	4th International Symposium on Advanced Processing and Manufacturing Technologies (APMT) for Structural and Multifunctional Materials and Systems
Symposium 9	Porous Ceramics: Novel Developments and Applications
Symposium 10	Thermal Management Materials and Technologies
Symposium 11	Advanced Sensor Technology, Developments and Applications

Focused Session 1 Geopolymers and other Inorganic Polymers
Focused Session 2 Global Mineral Resources for Strategic and Emerging
                   Technologies
Focused Session 3 Computational Design, Modeling, Simulation and
                   Characterization of Ceramics and Composites
Focused Session 4 Nanolaminated Ternary Carbides and Nitrides (MAX Phases)

The conference proceedings are published into 9 issues of the 2010 Ceramic Engineering and Science Proceedings (CESP); Volume 31, Issues 2–10, 2010 as outlined below:

- Mechanical Properties and Performance of Engineering Ceramics and Composites V, CESP Volume 31, Issue 2 (includes papers from Symposium 1)
- Advanced Ceramic Coatings and Interfaces V, Volume 31, Issue 3 (includes papers from Symposium 2)
- Advances in Solid Oxide Fuel Cells VI, CESP Volume 31, Issue 4 (includes papers from Symposium 3)
- Advances in Ceramic Armor VI, CESP Volume 31, Issue 5 (includes papers from Symposium 4)
- Advances in Bioceramics and Porous Ceramics III, CESP Volume 31, Issue 6 (includes papers from Symposia 5 and 9)
- Nanostructured Materials and Nanotechnology IV, CESP Volume 31, Issue 7 (includes papers from Symposium 7)
- Advanced Processing and Manufacturing Technologies for Structural and Multifunctional Materials IV, CESP Volume 31, Issue 8 (includes papers from Symposium 8)
- Advanced Materials for Sustainable Developments, CESP Volume 31, Issue 9 (includes papers from Symposia 6, 10, and 11)
- Strategic Materials and Computational Design, CESP Volume 31, Issue 10 (includes papers from Focused Sessions 1, 3 and 4)

The organization of the Daytona Beach meeting and the publication of these proceedings were possible thanks to the professional staff of ACerS and the tireless dedication of many ECD members. We would especially like to express our sincere thanks to the symposia organizers, session chairs, presenters and conference attendees, for their efforts and enthusiastic participation in the vibrant and cutting-edge conference.

ACerS and the ECD invite you to attend the 35th International Conference on Advanced Ceramics and Composites (http://www.ceramics.org/icacc-11) January 23–28, 2011 in Daytona Beach, Florida.

Sanjay Mathur and Tatsuki Ohji, Volume Editors
July 2010

Thermal and Environmental Barrier Coatings

KINETICS AND MECHANISM OF OXIDATION OF THE REINFORCED CARBON/CARBON ON THE SPACE SHUTTLE ORBITER

Nathan Jacobson and David Hull
NASA Glenn Research Center
Cleveland, OH 44135

James Cawley
Case Western Reserve University
Cleveland, OH 44106

Donald Curry
The Boeing Company
Houston, TX 77058

ABSTRACT

Reinforced carbon/carbon (RCC) protects the Space Shuttle Orbiter wing leading edge and nose cap from the heat of re-entry. The oxidation protection system is based on a SiC conversion coating. In this paper, fundamental laboratory oxidation studies from 600-1200°C on RCC fabric, fabric and matrix, and SiC-protected carbon/carbon are discussed. Although conducted under laboratory conditions, these studies reproduce the morphologies observed under flight conditions. Oxidation below ~900°C is reaction-controlled and is characterized by preferential attack of the matrix and thinning of the fibers. Further microscopy reveals that oxidation begins with preferential attack into the grooves of the fibers. At ~900°C, there is a transition to diffusion-controlled oxidation. This is characterized by more localized attack near the surface of the carbon/carbon. For SiC-coated RCC, cavities form at the base of cracks in the SiC. This morphology lends itself to a two-step diffusion model for carbon oxidation. Fluxes are considered in both the SiC channel and growing cavity. Experiments show good agreement with the model predictions. These studies show the criticality of a stable coating system with filled cracks to protect RCC.

INTRODUCTION

A unique series of design features and materials protect the Space Shuttle Orbiter from the heat of re-entry, which leads to temperatures in excess of 2800°F (1538°C) for several minutes in certain locations. The hottest parts of the Orbiter are the wing leading edge and nose cap. These are fabricated of a reinforced carbon/carbon (RCC) material which has performed remarkably well in nearly thirty years of service and well over 100 flights. Fig. 1 shows the location of the RCC panels on the Orbiter.

RCC was developed well over thirty years ago by Vought Aircraft (now Lockheed Missiles and Fire Control). [1,2] A diagram of this material is shown in Fig. 2 and an actual cross section is shown in Figs. 3(a) and (b). It is composed of a two-dimensional lay-up of carbon fabric. The fibers in this cloth are derived from a rayon precursor and they are about 10 μm in diameter with a fairly uniform microstructure.[3] Fig. 3(b) shows that the outer edges of the fibers exhibit crenulations or grooves, which are characteristic of wet-spun fibers.[4]

The two dimensional lay-up of cloth is repeatedly infiltrated with a liquid carbon precursors to fill porosity. However, it should be noted that not all voids are filled and, in fact, some of the additional processing steps require interconnected porosity. Fig. 3(a) shows a polished cross-section of RCC. Note the extensive porosity in the carbon/carbon substrate. This is due to incomplete compaction and shrinkage cracks on cooling.

Oxidation protection is, of course, a critical issue for RCC. The primary oxygen barrier is a conversion coating of SiC.[1,2] Due to the thermal expansion mismatch the SiC forms through-thickness cracks upon cooling from the processing temperature of ~1650°C. The entire piece is vacuum infiltrated with TEOS (tetra-ethyl orthosilicate), which decomposes on a mild heat treatment to SiO_2. Then the faces are coated with a glass sealant. This sodium silicate glass becomes fluid at high temperatures and effectively seals the cracks in the SiC. The sodium silicate contains SiC grit and chopped SiC fibers which act as a filler and limit frothing. The oxidation protection system is illustrated in the schematic in Fig. 2 and the actual cross-section in Fig. 3(a).

The glass system is remarkably effective and post-examination of specimens exposed to hot gases, when the glass is molten, show little oxidation. Nonetheless, there are conditions when the oxidation protection system is not operative. At low temperatures, when the glass is not fluid, cracks and fissures in the SiC coating can lead to oxidation of the carbon/carbon. At higher temperatures, the glass becomes depleted either due to vaporization or shear forces, and oxidation may occur. Vaporization, primarily as loss of the sodium component, becomes appreciable at temperatures greater than ~1200°C. The outer coating of sodium silicate has been on a regular refurbishment schedule to prevent this, but some oxidation has been observed on panels as the glass is depleted.

For this reason, RCC oxidation has been studied by several investigators.[3, 5-9] These studies have been conducted in arc-jets,[5,6] which most accurately reproduce the re-entry environment in regard to pressures, temperature, velocities and dissociated oxygen. However, arc-jet tests are complex and expensive and it is difficult to accurately control all variables. We have found that many of the morphological features observed on mission-exposed RCC can be reproduced in simple laboratory furnace tests. These laboratory tests involve diatomic oxygen at low flow rates and only can capture the temperature component of re-entry. Nonetheless, a good deal can be learned from these basic studies.

The reaction of carbon and oxygen at elevated temperatures is one of the most studied topics in high temperature oxidation.[10-14] The key steps in oxidation are given below.
1. Diffusion of oxygen to the carbon surface.
2. Chemical reaction at the surface.
3. Diffusion of the products CO and/or CO_2 away from the surface.
In general, carbon oxidation exhibits a low temperature (~400-900°C) region where the overall rate is controlled by the chemical reaction. In reaction-controlled oxidation, the rates are highly dependent on temperature and the type of carbon. Morphologies in this region indicate selective attack of specific microstructural features. At temperatures above ~900°C, the process is controlled by diffusion of oxygen to the sample or diffusion of the products away from the sample. In diffusion-controlled oxidation, the reaction rates exhibit only a weak dependence on temperature. Generally a more uniform reaction front is expected, as the rate of oxidation does not depend on the particular form of carbon in a localized region.

In this paper, the kinetics and mechanisms of RCC oxidation are reviewed. We discuss RCC with and without a SiC coating. The transition from reaction-control to diffusion-control is discussed both in terms of the Arrhenius plot and some important morphological changes. Additionally, a model for diffusion-controlled reaction is discussed.

EXPERIMENTAL

The starting material was the same type of RCC used on the Orbiter. Oxidation of fibers in the form of cloth and oxidation of the composite material with and without a conversion coating of SiC was studied. The RCC with a SiC coating was in the form of 1.91 cm diameter discs with all sides coated.

Fibers in the form of a cloth were oxidized in a thermogravimetric apparatus (TGA). The TGA used consisted of a vertical tube furnace with a 5 cm hot zone and microbalance. Cloth samples of ~0.30 g were placed in an alumina cup with slots cut in the bottom for gas flow. The carbon/carbon composites were prepared as cubes with ~0.5 cm sides and placed on an alumina platform with slots in the bottom. Tests were conducted in oxygen or air at a flow rate of 100 cc/min. Weight change and temperature data was collected with an automated collection system. Rates were taken as the instantaneous slope at 50% consumption, according to convention[14] for carbon materials. Oxidation rates were measured every 100 degrees between 500 and 1200°C.

Fibers and the actual carbon/carbon composite were mounted in epoxy and polished with diamond pastes for optical and electron-optical examination of both the as-fabricated and post-oxidation samples. Scanning electron microscopy was done with a Hitachi S4700 Cold Field Emission Scanning Electron Microscope (FE-SEM), equipped with an x-ray energy dispersive spectrometer (XEDS) for elemental analysis. A thin carbon film was evaporated onto the surface of mounted and polished samples to provide conductivity for the FE-SEM examination.

For the studies relating to the diffusion model, two sets of experiments were performed. The first set of experiments were done with 1.9 cm diameter disc of RCC with a SiC coating on all sides and a TEOS treatment. Slots of 0.53 and 1.02 mm width were made to the SiC/carbon-carbon interface. Oxidation treatments were done at 1200°C in a box furnace with static laboratory air. Oxidation damage was assessed by weight changes and optical microscopy of cross-sections. Image analysis software (Foveapro, Reindeer Graphics) was used to measure the cross sectional area of the oxidation cavity. The areas were approximated as a semi-circle and the radii extracted.

The second set of experiments were done with 1.9 cm diameter disc of RCC with a SiC coating on all sides and a TEOS treatment. Slots were not machined in these and oxidation occurred only through the natural craze cracks. The crack pattern was clearly revealed in a specimen by polishing a few hundred microns off the surface, as shown in Figs. 4(a) and (b). From this photo as well as cross sectional views, crack parameters such as crack length/unit area, crack width, and coating thickness were measured and used in the model. The specimens with natural craze cracks were oxidized in the TGA.

RESULTS AND DISCUSSION

A typical TGA plot is shown in Fig. 5(a) for the fibers as cloth at 900°C in air. In two hours the fibers were completely consumed. A typical TGA plot for the fibers and matrix material at 600°C in air is shown in Fig. 5(b) and is significantly different. It takes nearly fifty hours to consume the fiber and matrix at this lower temperature. The rapid weight loss at the beginning of the test is likely due to matrix oxidation, which often oxidizes faster than the fibers as it is less crystalline. This is consistent with microstructural observations.[3]

Fig. 6 is an Arrhenius plot of the oxidation rates of the fibers. These data show a characteristic change from a strongly temperature-dependent region below ~800°C to a weakly temperature-dependent region above ~800°C. The lower temperature region is thus attributed to reaction-control; whereas the higher temperature region is attributed to diffusion-control. Fig. 6 shows rates measured in air and oxygen. While there is not enough data to obtain an oxygen pressure dependence, the rates definitely decrease with oxygen pressure. Also note that the results of Halbig[16] for a T-300 carbon fiber are similar in the diffusion control region, but differ in the reaction-control region. The T-300 carbon fibers have outer layers of porosity which leads to diffusion control at lower temperatures.[11]

Reaction-control leads to a distinct morphology of attack. Fig. 7 is a micrograph of the fibers after 6.5 hrs at 600°C in air. Note the preferential attack in the form of fissures into the fibers. These fissures are not present in the as-fabricated material and are formed during oxidation. Attack is also observed in the form of thinning or denuding of the fiber diameter, leading to the smaller diameter fibers observed in Fig. 8(a).

Figs. 8(a) and (b) illustrate the reaction-controlled and diffusion-controlled oxidation in the RCC fibers and matrix, respectively. Note the reaction-controlled microstructure indicates oxidation throughout and thinner fibers. However the diffusion-controlled microstructure indicates a dense center, with oxidative attack near the outer edges. This is consistent with the observation of Halbig and Cawley on oxidized carbon fiber/SiC matrix composites.[17] Generally, the Arrhenius behavior and reaction-control to diffusion-control transition of RCC fibers and matrix material is very similar to the fibers alone.[3] The major difference is the matrix tends to oxidize faster than the fibers,[3] as shown in Fig. 5(b). However, as the matrix is heated to higher temperatures and graphitizes, it is likely that the matrix and fiber would begin to show similar oxidation rates.[8]

For SiC-protected RCC, the behavior is more complex. The transition from reaction-control to diffusion-control is easiest to observe microstructurally, as shown in the sequence in Fig. 9. In this sequence a small slot has been cut in the SiC coating to the carbon/carbon substrate to provide a clear path for oxygen to the carbon/carbon substrate. Note that at the lowest temperatures, there is only a small region around the slot which is oxidized, as evidenced by the thinner fibers in this region. As temperature is increased, a hemispherical cavity (in two dimensions) or half-cylinder trough (in three dimensions) forms. This type of attack is characteristic of diffusion-control oxidation where attack is uniform in all directions. Further, the geometry of such attack allows it to be readily modeled. These models have been discussed elsewhere,[9] but will be briefly summarized here.

In low oxygen potentials, carbon oxidation leads to CO(g). However, CO(g) and O_2(g) are thermodynamically incompatible. They will react to form the more thermochemically stable CO_2(g). For this reason, oxidation of carbon is generally treated in two steps:

At the carbon/gas interface: $C(s) + CO_2(g) = 2\ CO(g)$ [1]

At a distance away from the carbon/gas interface: $CO(g) + \frac{1}{2}\ O_2(g) = CO_2(g)$ [2]

The net reaction is the oxidation of C to CO(g), but separation to these two steps avoids the CO/O_2 incompatibility problem. Further, combustion studies of carbon show this secondary reaction front is located at a distance away from the solid surface.[18]

There are several studies in the literature of carbon oxidation through channels (slots or cracks) in an inert matrix.[5-13] The equations for this type of oxidation are well established and have been presented by these investigators. In this treatment, we include diffusion not only through the SiC channel, but also through the growing oxidation cavity. A schematic of this process is shown in Fig. 10.

As suggested by the microstructure, the diffusion equations for the channel are in rectangular coordinates and the growing cavity in polar coordinates. The equations for the flux in the channel are written as:

$$J_i = D_i^{eff}\left(\frac{\partial c_i}{\partial x}\right) + v_i^{ave} c_i$$ [3]

Here J_i is the flux of species i, D_i^{eff} is the effective diffusion coefficient of the species i, c_i is the concentration of species i, x is the distance along the channel, and v_i^{ave} is the average velocity of species i. The first term in this expression is the diffusive flux and the second term is the convective flux. The convective flux can be simplified since:

$$v_i^{ave} = \frac{\sum_i c_i v_i}{\sum_i c_i} = \frac{\sum_i J_i}{c_T}$$ [4]

Here c_T is the total of the concentration of species i. The boundary conditions are as follows:

At the base of the channel: $x = 0$: $\quad c_{CO} = c_{CO}^0 \quad c_{CO_2} = c_{CO_2}^0$

At the site of reaction [2]: $x = x_f$: $\quad c_{CO_2} = c_{CO_2}^* \quad c_{O_2} = c_{CO} = 0$ [5]

At the mouth of the channel: $x = L$: $\quad c_{O_2} = c_{O_2}^l \quad c_{CO_2} = 0$

As discussed, these equations have been solved analytically by several investigators.[7, 9, 19-24] The first step is to derive an expression for x_f/L, where L is the thickness of the coating. Then an expression for the flux of CO_2 is developed, which relates to the carbon consumption.

The oxidation model is further extended by adding the fluxes in the growing cavity. As noted, these are best written in polar coordinates.

$$J_{CO_2}^{tr} A' = -D_{CO_2} A' \left(\frac{\partial c_{CO_2}}{\partial r} \right) - \frac{A' c_{CO_2} J_{CO_2}^{tr}}{c_T}$$ [6]

Here $J_{CO_2}^{tr}$ is the flux of CO_2 in the growing 'trough' (see Fig. 8), A' is the area of exposed carbon for oxidation, and r is the radius of the trough. The flux of CO_2 at the base of the channel is then equated to the flux of CO_2 entering the trough. The resultant equation is solved for the concentration of CO_2 at r_1 (radius at time 0), $c_{CO_2}^{r_1}$, and this quantity is then inserted into the integrated form of equation [6]. Assuming a reaction probability of one, the flux of CO_2 can be related to both a radius of the growing trough and a weight loss. It is not possible to solve for trough radius or weight loss as a function of time. Nonetheless, the reverse can be done and the following equations are used to generate oxidation kinetics. The time, t, as function of trough radius, r_2, is given by:

$$t = \frac{M_{CO_2}}{M_c} \frac{\rho}{D_{CO_2} c_{CO_2}^{\cdot}} \left[\frac{\frac{r_2^2}{2} \ln(r_2) - \frac{r_2^2}{4} - \frac{r_2^2}{2} \ln r_1 + \frac{r_2^2}{2} \left(\frac{\pi x_f \left(c_T + c_{CO_2}^{\cdot} \right)}{r_1 c_T} \right) + \frac{r_1^2}{4} - }{\frac{r_1^2}{2} \left(\frac{\pi x_f \left(c_T + c_{CO_2}^{\cdot} \right)}{r_1 c_T} \right)} \right]$$ [7]

The time, t, as a function of weight loss, W, is given by:

$$t = \frac{M_{CO_2}}{M_c} \frac{\rho}{D_{CO_2} c_{CO_2}^{*}} \left[\frac{\left(\sqrt{\frac{2W}{\pi \rho l}} + r_1 \right)^2}{2} \ln \left(\sqrt{\frac{2W}{\pi \rho l}} + r_1 \right) - \frac{\left(\sqrt{\frac{2W}{\pi \rho l}} + r_1 \right)^2}{4} - \frac{\left(\sqrt{\frac{2W}{\pi \rho l}} + r_1 \right)^2}{2} \ln r_1 + \frac{\left(\sqrt{\frac{2W}{\pi \rho l}} + r_1 \right)^2}{2} \left(\frac{\pi x_f \left(c_T + c_{CO_2}^{*} \right)}{r_1 c_T} \right) + \frac{r_1^2}{4} - \frac{r_1^2}{2} \left(\frac{\pi x_f \left(c_T + c_{CO_2}^{*} \right)}{r_1 c_T} \right) \right]$$ [8]

Here M_i is the molecular weight of species i, D_i is the diffusivity of species i, $c_{CO_2}^{\cdot}$ is the concentration of CO_2 at position x_f, ρ is the density of carbon/carbon, l is the crack length, c_T is the total concentration of gaseous species in the crack, and r_1 is shown in Fig. 9(c). Although equation [8] is complex, it was found that the linear terms dominate and a linear weight loss term can be derived for comparison to experiment.

An important consideration is the change of the channel geometry with temperature. Two possible changes can occur. The craze cracks should ideally close as temperature increases. Clearly this does not occur, as oxidation is still observed. For the purposes of this approximation, the room temperature crack width is used. The second consideration is crack wall oxidation. Electron microscope observations indicated only very thin oxide films (<0.5 μm) are formed and most of the oxidation is internal oxidation of the porous SiC coating.[9] For this reason, oxidation crack width changes due to oxidation were neglected.

As discussed in the experimental section, two types of experiments were done to test the model. First slots were machined and radii measured from cross-sections, similar to Fig. 9(d). Fig. 11 compares experimentally measured radii from oxidation exposures to the model for two different slot widths. The solid line shows the model for diffusion only in the slot, ignoring the growing oxidation void. The dashed line shows the model for the coupled equations of diffusion in the slot and the void (equation [7]), and shows good agreement with the experimental data.

The second type of experiment was performed with the RCC being coated on all sides with SiC and oxidation through the naturally occurring craze cracks was measured. Specimens were characterized as described in the experimental section and oxidation exposures were done in a TGA with flowing air. These continuous weight-change measurements exhibited a weight gain for the first 1 hr or so of the oxidation run and then a smooth linear weight-loss. This weight-loss rate was determined and reported in Table I.

The results from the model are shown in Table I. Neither the craze cracks or the resultant oxidation cavities (Fig. 12) formed the uniform geometries observed with the machined cracks. A 'tortuosity' factor was not used to account for this non-ideality. Nonetheless, the agreement between the model and the experiments is quite reasonable suggesting the deviations from non-ideality are not great and/or tend offset each other.

Table I. Oxidation rates for SiC coated RCC specimens with natural craze cracks at 1200°C in air at 0.1 MPa.

Temperature, °C	Area of Carbon exposed craze cracks, mm^2	Measured weight loss mg/mm^2-hr	Calculated weight loss mg/mm^2-hr
1000	3.60 ± 0.923	26 ± 7	14.7
1100	3.96 ± 0.63	21 ± 6	13.6
1200	3.99 ± 0.132	30 ± 8	16.1
1300	3.97 ± 0.019	41 ± 9	16.8

SUMMARY AND CONCLUSIONS

Reinforced carbon/carbon has been successfully used on the Orbiter wing leading edge and nose cap for well over 100 missions. The structure of this remarkable material has been discussed. Oxidation is a primary concern for RCC as well as for other carbon/carbon structural materials. Basic laboratory studies on oxidation of the rayon-derived carbon fibers, carbon fibers and matrix, and SiC protected carbon/carbon have been discussed. Characteristic microstructures are shown for the reaction-controlled and diffusion-controlled regions. In the reaction-controlled region, oxidation begins with attack down fissures, which begin at the grooves or crenulations of the fibers. This is also accompanied with a thinning of the fibers. Diffusion-controlled oxidation leads to a more uniform attack, since all constituents are attacked evenly. In SiC-protected RCC, oxidation takes place in the form of a trough at the base of a crack. This type of geometry and process lends itself to modeling with a two-step carbon oxidation diffusion model. The model shows close agreement to laboratory data.

REFERENCES

[1] D. M. Curry, H. C. Scott, C. N. Webster, Materials Characteristics of Space Shuttle Reinforced Carbon-Carbon, *24th National SAMPE Symposium and Exhibition*, Volume 24, Book 1, 1979.

[2] D.M. Curry, J.W. Latchem, and G.B. Wisenhunt, in *21st AIAA Aerospace Sciences Meeting* (AIAA, 1983).

[3] N. Jacobson and D. Hull, Characterization and Oxidation of Rayon Fibers, in preparation for *Oxid. Met.*, 2010.

[4] D. D. Edie and R. J. Diefendorf, Carbon Fiber Manufacturing, *Carbon-Carbon Materials and Composites*, eds. J. D. Buckley and D. D. Edie (Noyes, Park Ridge, NJ, 1993), p.19.

[5] J. E. Medford, Prediction of Oxidation Performance of Reinforced Carbond-Carbon Material for Space Shuttle Leading Edges, Paper 75-730, *AIAA 10th Thermophysics Conference*, Denver, CO, May 27-29, 1975.

[6] J. E. Medford, Prediction of In-Depth Oxidation Distribution of Reinforced Carbon-Carbon Material for Space Shuttle Leading Edges, Paper 77-783, *AIAA 12th Thermophysics Conferencerence*, June 27-29, 1977.

[7] N. S. Jacobson, T. A. Leonhardt, D. M. Curry, and R. A. Rapp, Oxidative attack of carbon/carbon substrates through coating pinholes, *Carbon*, **37**, 411-419 (1999).

[8] N. S. Jacobson and D. M. Curry, Oxidation microstructure studies of reinforced carbon/carbon, *Carbon*, **44**, 1142-1150 (2006).

[9] N. S. Jacobson, D. J. Roth, R. W. Rauser, J. D. Cawley, and D. M. Curry, Oxidation through coating cracks of SiC-protected carbon/carbon, *Surface and Coatings Technology*, **203**, 372-383 (2008).

10. P. L. Walker, Jr., F. Rusinko, Jr., and L. G. Austin, Gas Reactions of Carbon, *Advances in Catalysis and Related Subjects*, Vol. XI, eds. P. D. Eley, P. W. Selwood, and P. B. Weisz Academic Press, New York, 1959, p. 133.

[11] K. S. Goto, K. H. Han, and G. R. St. Pierre, A Review on Oxidation Kinetics of Carbon Fiber/Carbon Matrix Composites at High Temperatures, *Transactions ISIJ*, **26**, 597-603 (1986).

[12] D. W. McKee, Oxidation Behavior and Protection of Carbon/Carbon Composites, *Carbon*, **25**, 551-557 (1987).

[11] I. M. K. Ismail, On the Reactivity, Structure, and Porosity of Carbon Fibers and Fabrics, *Carbon*, **29**, 777-792 (1991).

[12] P. Crocker and B. McEnaney, Oxidation and Fracture of a Woven 2D Carbon-Carbon Composite, *Carbon*, **29**, 881-885 (1991).

[13] J. Rodriguez-Mirasol, P. A. Thrower, and L. R. Radovic, On the Oxidation Resistance of Carbon-Carbon Composites: Importance of Fiber Structure for Composite Reactivity, *Carbon*, **33**, 545-554 (1995).

[14] W. H. Glime and J. D. Cawley, Oxidation of Carbon Fibers and Films in Ceramic Matrix Composites: A Weak Link Process, *Carbon*, **33**, 1053-1060 (1995).

[15] F. Lamouroux, X. Bourrat, R. Naslain, and J. Sevely, Structure/Oxidation Behavior Relationship in the Carbonaceous Constituents of 2D-C/PyC/SiC Composites, *Carbon*, **31**, 1273-1288 (1993).

[16] M. C. Halbig, J. D. Cawley, A. J. Eckel, and D. N. Brewer, Oxidation Kinetics and Stress Effects for the Oxidation of Continuous Carbon Fibers with Microcracked C/SiC Ceramic Matrix Composites, *J. of the Am. Ceram. Soc.*, **91**, 519-526 (2008).

[17] M. C. Halbig and J. D. Cawley, "Modeling the Oxidation Kinetics of Continuous Carbon Fibers in a Ceramic Matrix," NASA/TM—2000-209651, January 2000.

[18] A. M. Kanury, Introduction to Combustion Phenomena, p. 204, Gordon and Breach, New York, (1982).

[19] J. Bernstein and T.B. Koger, Carbon Film Oxidation Undercut Kinetics, *J. Electrochem. Soc.*, **135**, 2086-2090 (1988).

[20] E.L. Courtright, J.T. Prater, G.R. Holcomb, G.R. St. Pierre, and R.A. Rapp, Oxidation of Hafnium Carbide and Hafnium Carbide with Additions of Tantalum and Praseodymium, *Oxid. Met.*, **36**, 423-437 (1991).

[21] L. Filipuzzi, G. Camus, R. Naslain, and J. Thebault, Oxidation Mechanisms and Kinetics of 1D-SiC/C/SiC Composite Materials: I, An Experimental Approach, *J. Am. Ceram. Soc.*, **77**, 459-466 (1994).

[22] L. Filipuzzi and R. Naslain, Oxidation Mechanisms and Kinetics of 1D-SiC/C/SiC Composite Materials: II, Modeling, *J. Am. Ceram. Soc.* 77, 467-480 (1994).

[23] G.R. Holcomb, Countercurrent Gaseous Diffusion Model of Oxidation Through a Porous Coating, *Corrosion*, 52, 531-539 (1996).

[24] A. J. Eckel, J. D. Cawley, and T. A. Parthasarathy, Oxidation Kinetics of a Continuous Carbon Phase in a Nonreactive Matrix, *J. Am. Ceram. Soc.*, **78**, 972-980 (1995).

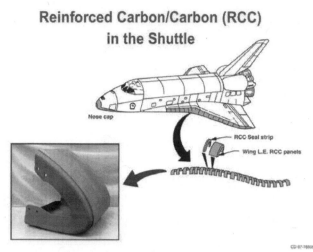

Figure 1. Location of RCC Panels on the Orbiter

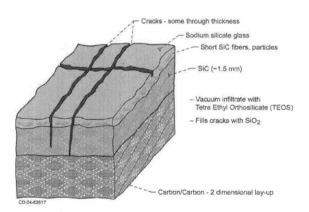

Figure 2. Schematic of RCC

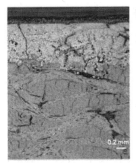

Glass Sealant

SiC Conversion Coating

Carbon/Carbon

Figure 3(a). Cross section of as-fabricated RCC showing carbon/carbon substrate, SiC protective coating, and glass sealant. Note the extensive porosity in the carbon/carbon substrate.

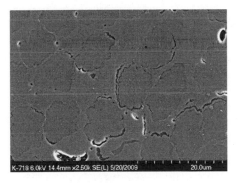

Figure 3(b). Cross section of as-fabricated RCC substrate. The ~10 μm round features with crenulations are the fibers and the material in between is the carbon matrix.

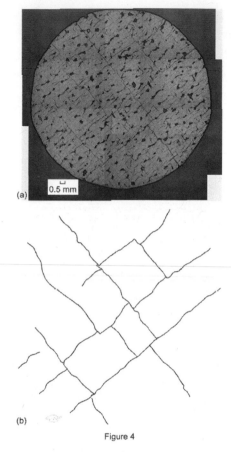

Figure 4

Figure 4. (a) RCC sample polished to reveal crack pattern (b) 'Skeleton' trace of that crack pattern. Reprinted from Reference (9) with permission from Elsevier

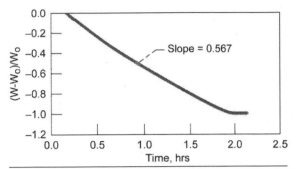

Figure 5(a). Representative TGA data for the rayon-derived fibers used in RCC at 900°C in air.

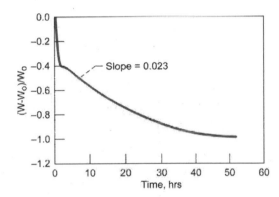

Figure 5(b). Representative TGA data for RCC (fibers and matrix) with no SiC coating at 600°C in air.

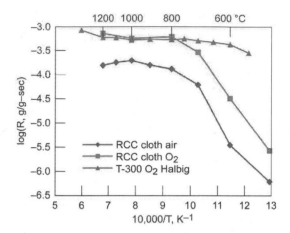

Figure 6. Arrhenius plot for oxidation of RCC cloth compared to T-300 carbon fibers.

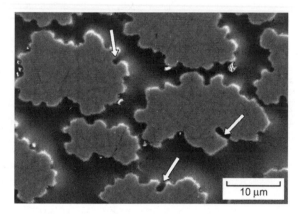

Figure 7. Initial fiber oxidation attack at 600°C for 6.5 hrs in air. The arrows point to the beginnings of oxidation attack at the grooves or crenulations.

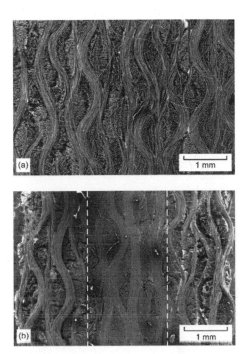

Figure 8. FE-SEM of cross-sections showing transition from reaction-control to diffusion-control. (a) Oxidation in air for 0.5 hr at 667 Pa and 800°C. This illustrates the uniform attack as predicted for reaction control. (b) Oxidation in air 0.5 hr at 667 Pa and 1100°C. This shows oxidation attack in the outer regions and a dense core, which is indicative of diffusion control. Both figures from Ref. 3.

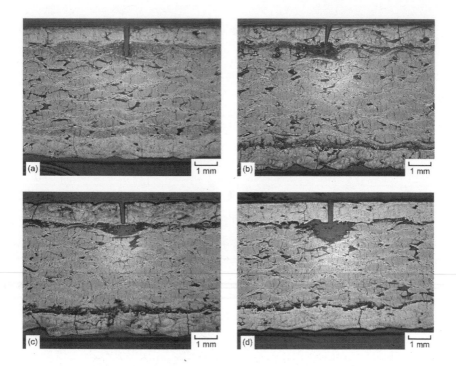

Figure 9. Images showing oxidation of SiC coated RCC specimens with a machined notch in the SiC coating in temperature sequence in air for 0.5 hr exposure time. Note the transition from reaction to diffusion control. (a) 650°C (b) 760°C (c) 870°C and (d) 983°C.

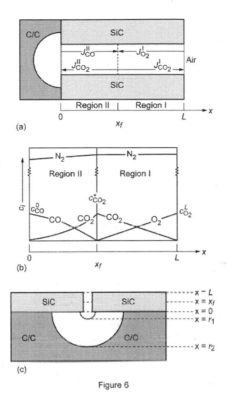

Figure 6

Figure 10. Schematic of oxidation process. (a) Opposing fluxes of gaseous products and reactants and growing cavity. (b) Relative concentrations in the channel. (c) Schematic of channel and cavity showing designation of boundary conditions in the equations. Reprinted from Reference (9) with permission from Elsevier.

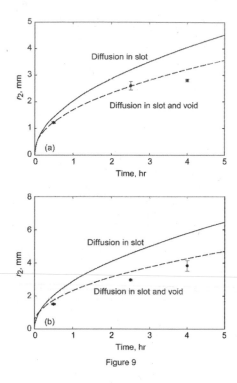

Figure 9

Figure 11. Comparison of experiment to model for (a) 0.53 width slot and (b) 1.02 mm width slot. Oxidation in air at 1 atm (0.1 MPa) at 1200°C. Circles are the experimental data. Reprinted from Reference (9) with permission from Elsevier.

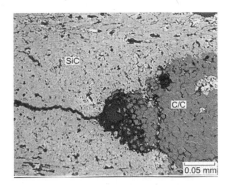

Figure 12. Diffusion-controlled oxidation through a crack. Oxidation at 1100°C, 669 Pa, for 1 hr.

THERMOMECHANICAL FATIGUE LIFE OF TBCS—EXPERIMENTAL AND MODELING ASPECTS

Sören Sjöström and Håkan Brodin
Siemens Industrial Turbomachinery AB
SE-612 83 FINSPÅNG, Sweden
And
Department of Management and Engineering, Linköping University
SE-581 83 LINKÖPING, Sweden

ABSTRACT
The fatigue life of APS TBC under TMF loading has been studied. Failure can be by spallation from convex surfaces, spallation from flat or nearly flat surfaces and spallation from sharp edges. The damage evolution leading to final failure has been studied experimentally, and based on the experimental observations, a fracture-mechanical model for the formation and growth of cracks in or near the thermally grown oxide and for the final failure of the TBC has been set up.

INTRODUCTION
The concept of thermal barrier coatings (TBCs) is well-known among gas-turbine designers. The fundamental idea of a TBC is that an insulating ceramic top coat (TC) is used for thermal protection of components which are exposed to hot gas (*e.g.*, combustor linings, turbine guide vanes and turbine blades). Between the substrate and the TC there is generally a bond coat (BC), which is a metallic alloy with double purposes. On one hand, it improves the adhesion between the substrate and the TC, and on the other hand it protects the substrate from high-temperature oxidation by itself oxidising into oxides with less unfavourable properties than oxides forming directly on the substrate would have had.

Depending on the application method there exist two main types of TBC, namely air-plasma-sprayed (APS) and electron-beam physical-vapour-deposited (EB-PVD). EB-PVD TBCs are mainly used in aeroengine turbine vanes and blades, while larger components, such as burner liners and turbine vanes and blades in stationary gas-turbine engines mainly use APS TBC. The TMF properties are quite different for the two types, and this article is therefore limited to the analysis of APS TBC.

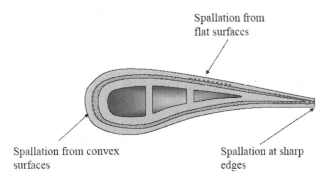

Spallation from
flat surfaces

Spallation from convex
surfaces

Spallation at sharp
edges

Figure 1 Basic failure mechanisms

BASIC FAILURE MECHANISMS

The multilayer structure of the TBC, in which, for instance, elastic, thermal expansion and fracture properties vary between the layers, makes it sensitive to thermal cycling and impact loads. Further, the oxide forming basically at the TC/BC interface has generally less favourable adhesion properties and also introduces internal stress by volume increase, which further threatens the integrity of the TBC aggregate after long-term high-temperature exposure and oxidation.

Fig. 1 shows an overview of the basic failure mechanisms that threaten an APS TBC. These are
1. Spallation on convex surfaces
2. Spallation on flat (or nearly flat) surfaces
3. Spallation at sharp edges
These basic failure mechanisms will now be treated in some detail.

SPALLATION ON CONVEX SURFACES

It is easy to show that cooling down to low temperature after a not too short time spent at high temperature will produce a tensile 'transverse' stress $\sigma_\perp$ over the TC/BC and BC/substrate interfaces. $\sigma_\perp$ can quickly become very large, and since the capacity of the interface to bear transverse stress is low, this is a generally problematic situation. A simple model, using the basic assumptions that

- at the end of the high-temperature hold time, the stresses near the interface are completely relaxed,
- the stresses arising during the cooling down are large enough to make the whole TC flow plastically in compression, and
- the plastic flow of the TC is non-hardening with yield stress Y^-

leads to the following equation of $\sigma_\perp$ at the end of the cooling down:

$$\sigma_\perp = Y^- \cdot \ln \frac{\rho}{\rho + d_{TC}} \tag{1}$$

Assuming that $Y^- = -100 \cdot 10^6$ Pa, for a thin TBC ($d_{TC} = 300 \cdot 10^{-6}$ m), $(\sigma_\perp)_{\text{allowed}} = 5 \cdot 10^6$ Pa therefore gives a minimum allowed radius of curvature $\rho_{\min} \approx 10$ mm.

SPALLATION ON FLAT (OR NEARLY FLAT) SURFACES. GENERAL DESCRIPTION OF THE MECHANISM

This failure mechanism has been much more difficult to understand, and it can well be argued that it is still not completely understood. On the macro scale, the sequence of events is that

- the adhesion between layers is lost over a sufficiently large area,
- the coating (which has lower thermal expansion than the substrate) will get into strong compressive stress at low temperature (after long high-temperature service) and therefore buckles,
- breaks up at the edge of the buckling area and
- breaks away.

What needs an explanation is therefore the mechanism behind the loss of adhesion over a sufficiently large area that starts the process.

In pursuing this, we need a physically sound model, relying on experimental observations as well as on a fracture-mechanical background.

SPALLATION ON FLAT (OR NEARLY FLAT) SURFACES. EARLY MODELS

In 1984, Miller et al.[1] came up with the idea of using a Coffin-Manson equation for the number N_{fN} of cycles left to to failure,

$$N_{fN} - (\frac{\varepsilon_{eN}}{\varepsilon_f})^b ,\qquad (2)$$

or, stated differently, for the damage D_N per cycle

$$D_N = \frac{1}{N_{fN}} = (\frac{\varepsilon_{eN}}{\varepsilon_f})^{-b} .\qquad (3)$$

In these equations, ε_f was a critical strain, N a cycle counter and b a negative exponent. The new idea was that a strain range ε_{eN} was to be used which depended on the actual oxide weight gain w_N:

$$\varepsilon_{eN} = \varepsilon_f \left[(1 - \frac{\varepsilon}{\varepsilon_f})(\frac{w_N}{w_c})^m + \frac{\varepsilon}{\varepsilon_f} \right] ,\qquad (4)$$

where w_c is a critical oxide weight gain. By this, a cycle after long high-temperature exposure would be more severe than an equal cycle early in the component's history. The failure criterion was stated as

$$\sum_{N=1}^{N_f} D_N = \sum_{N=1}^{N_f} \left[(1 - \frac{\varepsilon_r}{\varepsilon_f})(\frac{w_N}{w_c})^m + \frac{\varepsilon_r}{\varepsilon_f} \right]^{-b} = 1 \qquad (5)$$

In order to get a better understanding of the actual mechanics behind the spallation failure, FE computations of the stress state in the neighbourhood of the TC/BC interface were also performed by Chang and Phucharoen[2]. These computations showed that the thermally grown oxide (TGO) was likely to play an important role in the failure process by contributing to the formation of a strong state of vertical stress at or near the interface.

SPALLATION ON FLAT (OR NEARLY FLAT) SURFACES. FAILURE MODEL BASING ON FRACTURE-MECHANICAL MECHANISMS

Jinnestrand and Sjöström[3] made a 3D FEM analysis of a least representative cell of a TBC aggregate, see Fig. 2. The stress state was analysed for different thicknesses of TGO formed in the BC/TC interface. Substrate, BC and TC were modelled as creeping materials, and the volume growth of the TGO was modelled as anisotropic. The load cycles applied are shown in Fig. 3; different oxide thicknesses were analysed by interrupting the intermediate (oxidation) cycle after corresponding times. Fig. 4 shows that a region of tensile vertical stress exists at the tops of the wavy interface for the un-oxidised case and that this region of tensile vertical stress moves down along the flanks of the interface contour as the oxide thickness increases. Assuming that horizontal crack embryos exist in or near the interface at the contour tops at the start of the cycling history, it is therefore very likely that these can continue to grow along the flanks throughout the history of cycling and TGO growth.

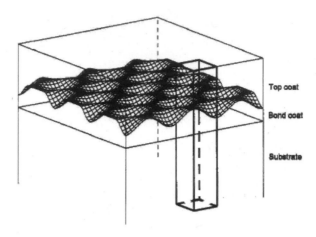

Figure 2   Least representative cell of a TBC aggregate

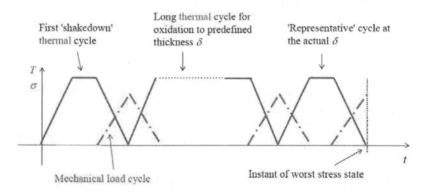

Figure 3   Load cycle applied in model of least representative cell

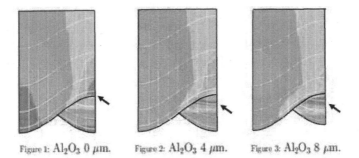

Figure 1: Al$_2$O$_3$ 0 μm.　　Figure 2: Al$_2$O$_3$ 4 μm.　　Figure 3: Al$_2$O$_3$ 8 μm.

Figure 4   Level plot of vertical stress on TC of least representative cell. Arrows mark points of maximum tensile stress (in the 0 μm case at the interface profile top, in the 4μm and 8 μm cases on the back flank). The dark areas near arrowheads have tensile stress > 100 MPa.

Thermal cycling tests

With this in mind, a series of thermal cycling tests was performed in a furnace, see Fig. 5. Test rig and test specimen data are listed in Tables 1 and 2. Specimens were taken out at predefined numbers of cycles, cut and analysed for horizontal cracks at or near the TC/BC interface. Fig. 6 shows the crack situation in four specimens taken out after different numbers of cycles. The vertically projected length of cracked interface was defined as a damage D:

$$D = \frac{\sum' l_i}{L} . \tag{6}$$

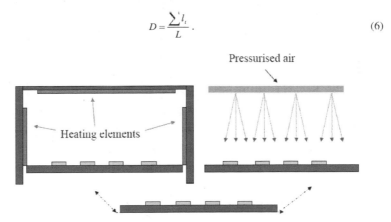

Figure 5   Rig used in the TCF testing. Cycling by moving tray carrying test specimens between furnace bottom and pressurised air cooling position

Table 1   Rig data

Rig data
T_{max} = 1100 °C
T_{min} = 100 °C
dT/dt = 1.7 °C/s
Dwell time at T_{max} 3600 s

Table 2   Specimen data

Specimen data
Substrate: Haynes 230, thickness d_{H230}=5 mm
BC: NiCoCrAlY, thickness d_{BC}=0.15 mm
TC: YSZ Zr2O3, thickness d_{TC}=0.3 mm
L×b×h: 30mm×50mm×5.45mm

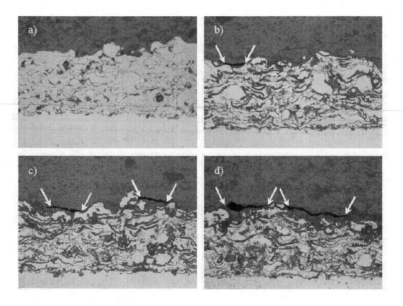

Figure 6   Development of cracks.
a) specimen taken out early for inspection, …, d) specimen taken out near failure.
Arrows mark crack ends.

When plotted against number of cycles, the development of D shows four typical stages, namely

I)   a short period of slow growth at the beginning of the life,

II)   a short period of strong growth,

III)   a period of slow growth at a level of D = 0.7 à 0.75, and

IV)   the final failure: BC/TC sufficiently weakened to allow buckling of TC, leading to spalling over a larger surface. This typically takes place at $D = D_c ≈ 0.85$.

See Fig. 7.

A closer examination of the details of the crack positions and geometries show that most cracks run in the TGO (which, in turn, forms in the TC/BC interface) or in the TC immediately above the TGO. The oxide itself was in this case shown to be almost entirely Al_2O_3 with only minor amounts of Cr, Co and Ni oxides.

Further details on the testing and modelling work can be found in Jinnestrand and Brodin[4] and Brodin and Eskner[5].

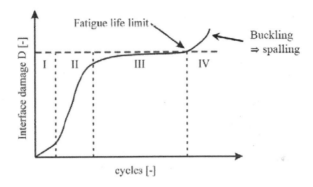

Figure 7   Measured damage evolution in APS TBC

Fracture-mechanical modelling

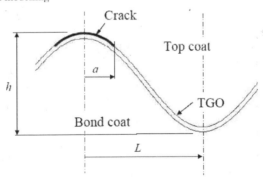

Figure 8   Idealised interface / crack geometry for the FM model

In parallel with the described experimental investigation, a fracture-mechanical model has been set up. The model bases on an idealised (sinusoidal) BC/TC interface profile, in which the TGO is assumed to grow. See Fig. 8. Equal cracks grow symmetrically from all profile tops, *i.e.*, when $D = a/L$ = 1.0, cracks extend along the whole interface, leading to complete failure. Referring back to the introductory explanation of the spallation mechanism, we will probably have a sufficient weakening of the interface already at some $D < 1.0$ for the final buckling of the TC to occur, and the actual critical damage D_c will therefore be < 1.0. Further essential model assumptions:

- Plane strain.
- The TBC system is assumed stress-free at the end of the high-temperature part of the cycle (high temperature gives rapid stress relaxation and relaxation of the misfit stress created by the TGO growth).
- The maximum damage-driving energy release rate G and stress intensity factors K_I and K_{II} appear after cooling down from the maximum temperature (main reason for this: low thermal expansion of the TC).
- The material behaviour of all components of the TBC aggregate during the cooling down from maximum temperature is assumed to be linearly elastic (the temperature decreases rapidly, and the time spent at high temperature during the cooling history is therefore not long enough to allow creep deformation to take place).
- G, K_I and K_{II} are computed by a virtual crack extension method and by interface crack theory.

The assumptions regarding material behaviour and stress state have been done in order to be able to focus on the fracture-mechanical analysis of the cooling cycle. The assumption of a stress-free state at the end of the hot period seems to be in line with some others' conclusions, whereas the assumption about linearly elastic cooling process is under debate. See, for instance Pindera et al.[6], Rösler et al.[7], Bäker et al.[8] and Busso et al.[9]

The computations have been performed for the same TBC and the same cycle as in the furnace test (i.e., substrate , BC and TC thicknesses, mechanical and thermal properties and thermal cycle were identical, cf. Tables 1 through 3). Interface profile height/half-period ratios used are

- $h/L =10/70, 20/70, 30/70$ and $50/70$.

For each of these, four oxide thicknesses

- $\delta_{TGO} = 0, 4\mu m, 6\mu m$ and $8\mu m$

have been analysed, and for each of these 16 combinations of h/L and δ_{TGO},

- a sequence of continuously increasing crack lengths corresponding to damage $D_m = a/L$ in the interval $0.0 < D_m < 1.0$

have been computed.

Table 3. Mechanical and thermal data used in the computations

	E (Pa)	ν	α (K^{-1})
TC	$48.8 \cdot 10^9$	0.25	$7.6 \cdot 10^{-6}$
TGO	$317 \cdot 10^9$	0.26	$8.0 \cdot 10^{-6}$
BC	$203 \cdot 10^9$	0.32	$12.0 \cdot 10^{-6}$
Substrate	$213 \cdot 10^9$	0.29	$12.2 \cdot 10^{-6}$

The results are shown in Fig. 9. The general tendency is that thicker TGO gives larger G, K_I and K_{II} . The influence of profile height ratio h/L will be commented later, but it is worth noticing that for high h/L, K_I increases rapidly to an early maximum, from which it then rapidly decreases to 0 and becomes negative (actually meaning, of course, that the crack-opening loading is replaced by contact forces on the crack flanks).

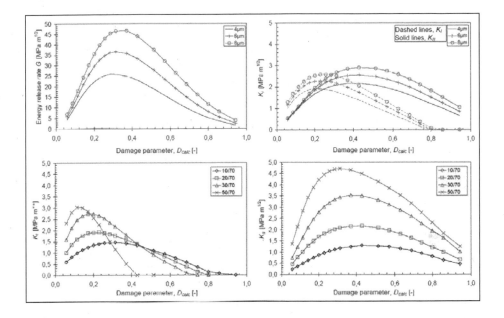

Figure 9   Energy release rate G and stress intensity factors K_I and K_{II} computed for the TCF test. The top two graphs are for $h/L = 30/70$, and the lower two graphs are for TGO thickness $\delta = 6$ μm.

TMF fatigue life model. Paris-law approach

Basing on the fracture-mechanical data from above, a Paris-law approach has been used for setting up a TMF fatigue life model. Noting that by previous sections, the crack length a can be formally replaced by the damage D,

$$D = \frac{a}{L} \Rightarrow a = DL, \tag{7}$$

Paris' law can be formally written as a damage evolution law,

$$\frac{dD}{dN} = C(\lambda \Delta G)^n \tag{8}$$

From the fracture-mechanical computations, we have

$$\Delta G = \Delta G(N,t) = f_G \left[\delta_{\text{TGO}}(t), D(N,t), h/L, \Delta \varepsilon_{\text{mech}}(N), \Delta T(N) \right] \tag{9}$$

$$\Delta K_I = \Delta K_I(N,t) = f_{K_I} \left[\delta_{\text{TGO}}(t), D(N,t), h/L, \Delta \varepsilon_{\text{mech}}(N), \Delta T(N) \right] \tag{10}$$

$$\Delta K_{II} = \Delta K_{II}(N,t) = f_{K_{II}}\left[\delta_{TGO}(t), D(N,t), h/L, \Delta\varepsilon_{mech}(N), \Delta T(N)\right] \tag{11}$$

λ in Eq. (8) is a factor to handle mixed-mode interface fatigue crack growth. An expression for this has been set up by Hutchinson and Suo[10]:

$$\lambda = 1 - (1 - \lambda_0) \cdot \frac{2}{\pi} \arctan\left(\frac{\Delta K_{II}}{\Delta K_I}\right) \tag{12}$$

(Note that $\Delta K_{II}/\Delta K_I \to 0 \Rightarrow \lambda \to 1$ and $\Delta K_{II}/\Delta K_I \to \infty \Rightarrow \lambda \to \lambda_0$.) Eqs. (8) through (11) can now be used for integrating the damage growth:

$$D(N,t) = D_0 + \int_0^N \frac{dD}{dN}\,dN = D_0 + C\int_0^N (\lambda\Delta G)^n \,dN, \tag{13}$$

The cycle N_f for which D reaches the value D_c gives the number of cycles to failure. The model will in the sequel be called the FM model.

Correlation between tests and FM model

It is now interesting to see if the FM model can describe the experimentally observed damage evolution and failure behaviour.

The FM model has therefore been run with exactly the same thermal cycle, elasticity data and thermal expansion data as the experiments. Parameter optimisation involving parameters C, λ_0 and n in Eq. (7) showed that the FM model could describe the experimental damage evolution closely, see Fig. 10. The calibrated parameter values are listed in Table 4.

Table 4. Calibrated parameters

Calibrated data
$D_0 = 0.05$
$C = 58.9 \cdot 10^{-6}$
$\lambda_0 = 0.1$
$n = 1.88$

Applicability of the FM model to other situations. 1) Out-of-phase (OP) and in-phase (IP) TMF situations.

In the measurements, buckling and spalling of the TC appeared for $D=D_c \approx 0.85$ (cf. limit entered in Fig. 7)

We have run experiments and FM model for the same material combination as before but with IP and OP TMF loads. Data used for the FM model were also the same as those used before, Results are shown in Fig. 11.

It can be concluded that in the OP TMF case, FM model results come acceptably close to the experimental ones, whereas the IP TMF case gives considerable deviation. The reasons of this must be further investigated, but the basic assumptions of linear elasticity and most dangerous part of the cycle can certainly be questioned in the IP case, where the crack-opening part of the load cycle occurs at high temperature. Microstructure examinations also reveal that other types of cracks appear, for instance vertical cracks in the top coat. The FM model as described above may therefore not be directly applicable in IP TMF situations.

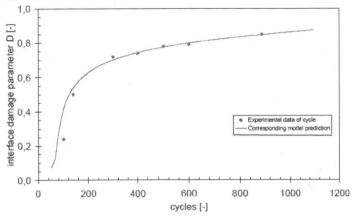

Fig 10 Comparison between measured and computed damage evolution (after parameter oprimisation).

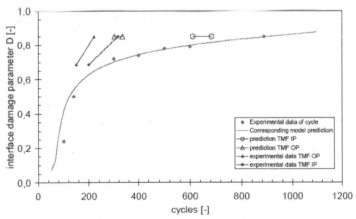

Fig 11 Comparison between measured and computed damage evolution for IP and OP TMF cycles

Applicability of the FM model to other situations. 2) Thicker TBC.

For a 1.5 mm TBC, measured damage evolution is shown in Fig. 12, and FM model results are shown in Fig. 13. The FM model gives somewhat faster damage growth than was measured. It is worth noting that cracks kinking out of the TGO into the top coat are more frequent than in the thin TBC case, meaning again that the damage mechanism assumed in the FM model (and basing primarily on observations for the 300 μm TBC) may not be completely relevant.

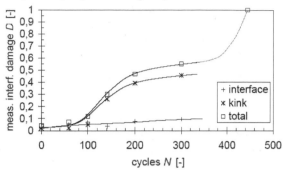

Figure 12   Measured damage evolution in 1.5 mm TBC

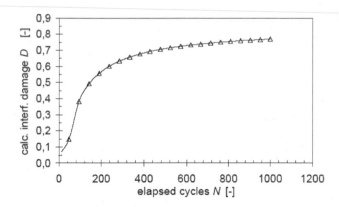

Figure 13   Computed damage evolution in 1.5 mm TBC

FM model damage evolution for different interface roughnesses
        We have finally compared TBC spallation damage evolution computed by the FM model with measured damage evolution in cases of different interface roughness. Fig. 14 shows this comparison for the measured N_f in the $R_a = 6$ μm and $R_a = 12$ μm cases, which correspond roughly to the $h/L = 20/70$ and $h/L = 30/70$ cases of the FM computations, respectively. The parameter optimisation (see above) was done for the 30/70 case, and it can be seen that the FM computations do not predict the measured N_f for the $R_a = 6$ mm case very well. One explanation can be found in Fig. 9, where it can be seen that for $h/L = 30/70$, K_I vanishes at $D \approx 0.75$ and is in our calibration runs assumed to be $= 0$ for

D > 0.75. A better assumption would probably have been to compute the actual compressive contact force F_N on the crack flanks for D > 0.75 in order to reduce K_{II} by the friction force caused by this F_N. A simple approximation of this method could be to continue to compute K_I even after it has become negative and reduce K_{II} in the following way:

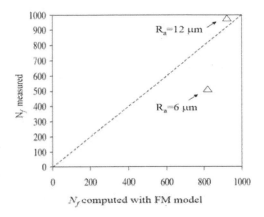

Fig 14 Comparison between measured and computed N_f for interface roughnesses $R_a = 6\mu m$ (corresponding to $h/L = 20/70$) and $R_a = 12 \ \mu m$ (corresponding to $h/L = 30/70$)

$$K_{II}^{reduced} = K_{II} + \mu K_I \quad \text{for } K_I < 0,\tag{14}$$

where $\mu > 0$ is the friction coefficient. The neglection of this reduction of crack growth rate due to crack flank friction for large D values in the $h/L = 30/70$ case lead us into too high computed damage evolution rates, which disturbed our calibration and lead us into too low C (cf. Eq. (8)). As a consequence, the calibrated parameters give too low computed damage evolution rates in, e.g., the $h/L = 20/70$ case, where crack flanks in contact ('negative' K_I) never appear (cf. Fig. 9).

A general remark on the choice of D_c

The 'regular' damage growth (by interface crack growth) is assumed to continue until the final buckling starts. We have therefore chosen to define the life as the end of the 'plateau' in the damage growth, e.g., in our main calibration case (cf. Figs. 7 and 10) at $D = D_c = 0.85$. As is seen by, e.g., a comparison with the thick-TBC case shown in Fig.12, this cannot be regarded as a universal value. Instead, D_c must be adjusted to parameters such as top coat thickness, oxide growth behaviour, oxide composition and also interface roughness.

SPALLATION AT SHARP EDGES

It is well-known that at sharp coating edges we will have stress singularities, *i.e.*,

$$\sigma_{ij} = K_s \cdot f_{ij}(\theta) \cdot r^{-\omega_s} \tag{15}$$

(see Fig. 15). It is therefore reasonable to suspect that the coating edge will be a source of early fatigue damage, and it is important to design the coating as carefully as possible to avoid this. One possible way would be to introduce a chamfered end, by which the order of the singularity would be reduced and the fatigue risk would be lowered.

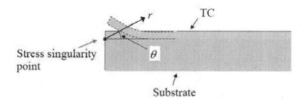

Figure 15   Stress singularity point

The mathematics behind the singularity is rather well developed, and for an unchamfered TC end on a TBC with representative material properties, the computed stress singularity order will be $\omega_s = 0.10$ for the case in Fig. 15 (in which the TC ends at the end of the substrate) and $\omega_s = 0.36$ for the case of a TC ending on a continuous substrate. It is, however, not easy to compute and understand the generalised stress intensity factor K_s in Eq. (15), and we have therefore set up a tentative model in order to analyse the fatigue spallation risk of a coating end.

Theory of critical distances

The theory of critical distances has originally been set up to handle fatigue situations for components with sharp notches. See, for instance, Susmel[11]. The case of a coating end can, however, be considered as a (very) sharp notch, and the theory can then be described in the following way:
Set up an 'equivalent fatigue risk evaluation stress':

$$\Delta\sigma_{\text{eff}} = \frac{1}{2L} \int_0^{2L} \Delta\tau_{xy}(\theta=0,x)dx \tag{16}$$

In Eq. (15),

$$L = \frac{1}{\pi}\left(\frac{\Delta K_{\text{th}}}{\Delta\sigma_0}\right)^2, \tag{17}$$

$\Delta\tau_{xy}$ is the actual shear stress range due to the temperature cycle, ΔK_{th} is the fatigue threshold, and $\Delta\sigma_0$ is the fatigue limit under uniaxial cyclic loading.
The stress is solved by a FEM model of the coating end geometry, see *e.g.* Fig. 16. $\Delta\tau_{xy}$ is shown for different chamfer angles $\phi^{(1)}$ in Fig. 17, and the corresponding computed $\Delta\sigma_{\text{eff}}$ are shown in Fig. 18.

The results are somewhat surprising, since they imply that no reduction in the risk of early end spallation is achieved for chamfer angles $\phi^{(1)} > 60°$ and that one must go down to at least $\phi^{(1)} = 45°$ to get a noticeable improvement.

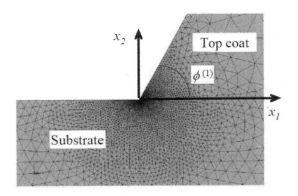

Figure 16 FEM model used in analysis of coating edge singularity. Case of TC end on continuous substrate.

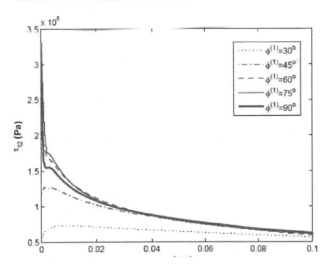

Figure 17 τ_{12} for different chamfer angles $\phi^{(1)}$

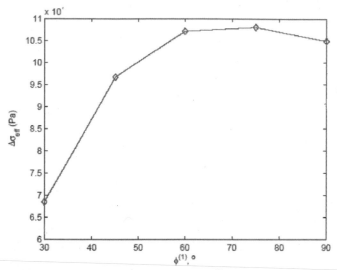

Figure 18 $\Delta\sigma_{eff}$ for different chamfer angles $\phi^{(1)}$

CONCLUSIONS

Damage growth in an APS TBC has been measured during a thermal-cycling testing. The results have been used for setting up a fracture-mechanics-based model for the damage evolution and final spallation failure of the APS TBC. The model has been tested for different applications, and it has been found to handle at least thin APS TBC (i.e., TC thickness 300 μm) reasonably well. Reasons of deviations have been identified and will be treated in the future. IP TMF and thick APS TBC (i.e., TC thickness 1.5 mm) show at least partially different damage mechanisms and must therefore become the objects of more thorough studies.

ACKNOWLEDGMENTS

In addition to Siemens Industrial Turbomachinery, the project work has been sponsored by the Swedish Energy Agency through grants
- KME-706
- KME-106 and
- Turbopower project No. 8

REFERENCES

[1]R.A. Miller, P. Argarwal, E.C. Duderstadt, Life model of atmospheric and low pressure plasma-sprayed thermal-barrier coating, *Ceram. Eng. Sci. Proc.*, **5**, 470-478 (1984)
[2]G.C. Chang, W. Phucharoen, Behavior of thermal barrier coatings for advanced gas turbine blades, *Surf. & Coat. Techn.*, **30**, 33-28, (1987)
[3]M. Jinnestrand, S. Sjöström, Investigation by 3D FE simulations of delaminatiuon crack initiation in TBC caused by alumina growth, *Surf. & Coat. Techn.*, **135**, 188-195 (2001)

[4]M. Jinnestrand, H. Brodin, Crack initiation and propagation in air plasma sprayed thermal barrier coatings, testing and mathematical modelling of low cycle fatigue behaviour, *Mater. Sci. Eng.*, **A 379**, 45-57 (2004)

[5]H. Brodin, M. Eskner, The influence of oxidation on mechanical and fracture behaviour of an air plasma-sprayed NiCoCrAlY bindcoat, *Surf. & Coat Techn.*, **187**, 113-121 (2004)

[6]M.-J. Pindera, J. Aboudi, S.M. Arnold, The effect of interface roughness and oxide film thickness on the inelastic response of thermal barrier coatings to thermal cycling, *Mater. Sci. Eng.*, **A 284**, 158-175 (2000)

[7]J. Rösler, M. Bäker, K. Aufzug, A parametric study of the stress state of thermal barrier coatings. Part I: creep relaxation, *Acta Mater.*, **52**, 4809-4817 (2004)

[8]M. Bäker, J. Rösler, G. Heinze, A parametric study of the stress state of thermal barrier coatings. Part II: cooling dtresses, *Acta Mater.*, **53**, 469-476 (2005)

[9]E.P Busso, Z.Q. Qian, M.P. Taylor, H.E. Evans, The influence of bondcoat and topcoat mechanical properties on stress development in thermal barrier coating systems, *Acta Mater.*, **57**, 2349-2361 (2009)

[10]J.W. Hutchinson, Z. Suo, Mixed mode cracking in layered materials, *Adv. Appl. Mech.*, **29**, 63-187 (1990)

[11] L. Susmel, The theory of critical distances: a review of its applications in fatigue, *Eng. Fract. Mech.*, **75**, 1706-1724 (2008)

OXIDATION-INDUCED STRESSES IN THERMAL BARRIER COATING SYSTEMS

Hugh E. Evans
School of Metallurgy and Materials
The University of Birmingham
Birmingham, B15 2TT, UK

ABSTRACT

Thermal barrier coating systems offer oxidation protection to alloy substrates under heat flux conditions. They consist of an yttria-stabilised zirconia top coat, of low thermal conductivity, bonded to the alloy substrate by an alumina-forming metallic layer. These systems develop considerable strain energy during cooling and are mechanically unstable. The failure by spallation of the top coat is a life-limiting event but its prediction has proved a difficult problem. The growth of the alumina layer on the bond coat surface has been implicated in the failure of the TBC system but complete understanding remains elusive. In this paper, finite-element computations are presented of the out-of-plane stresses that develop both isothermally and during cooling, as a result of oxide growth. A novel feature of the calculations is that oxide thickening and the associated volume expansion are modelled within the computation. It is shown that continuity strains can lead to the development of out-of-plane stresses at the base of the top coat at the oxidation temperature. These are associated with bond coat protuberances and their magnitude increases with surface roughness. The situation will be exacerbated if Al depletion is sufficient to trigger chemical failure and the formation of faster-growing Cr,Ni-rich oxides at bond coat protuberances. In this case, large (> 0.5 GPa) out of plane tensile stresses can develop within the top coat at the test temperature for realistic surface roughnesses.

INTRODUCTION

Thermal Barrier Coatings (TBC) systems consist of a thermally-insulating yttria stabilised zirconia (YSZ) top coat attached to the superalloy substrate by means of an oxidation-resistant bond coat. This system can develop a large temperature drop (e.g. 100-150°C) through the ceramic top coat under heat flux conditions and permit the alloy substrate to operate at significantly lower temperatures than if uncoated. It can be appreciated that TBC systems have the potential of offering an increase in the efficiency of both aero-engine and land-based gas turbine engines by increasing the inlet temperature and reducing the amount of cooling air required in high-temperature components. However, the commercial potential of these coating systems has not been realised fully due to the lack of understanding and thus control of the failure mechanisms, e.g. spallation of the outer ceramic layer. Failure of the TBC system through spallation of the ceramic under these circumstances would lead to excessively rapid oxidation of the bond coat and component[1-4].

The YSZ top coat will permit oxygen ingress, either molecularly or by solid state diffusion, but oxidation protection is provided to the alloy by the bond coat. All current production bond coats are designed to form a protective layer of alumina and are enriched in aluminium. Chromia-forming bond coats within TBC systems have recently been proposed[5] but these are intended for use under low-temperature hot corrosion conditions. Alumina-forming bond coats are designed to provide oxidation protection at high-temperature and also to establish a sufficient aluminium reservoir so that the protective alumina layer can be maintained. Current bond coats are either of overlay MCrAlY type (where M is Ni, Co, Fe or a combination of these) or they may be diffusion coatings produced by platinising and/or aluminising the outer surface of the nickel based superalloy substrate.

The formation and growth of the alumina layer is a selective oxidation process which preferentially consumes aluminium from the bond coat. The conversion to the oxide results in a volume expansion which occurs within the mechanically constrained location between the bond coat and top coat. This volume expansion results in the development of continuity strains and associated stresses within the ceramic top coat[6,7] on (typical) non-planar bond coat surfaces. The situation is exacerbated should fast-growing, non-protective oxides form locally as a result of localised aluminium depletion within the bond coat. These aspects will be considered in this paper as well as the influence of thermal strains produced during temperature changes. Initially, though, consideration will be given to the inherently mechanically unstable nature of a TBC system.

STRAIN ENERGY CONSIDERATIONS

The spallation of a protective oxide layer can be expected to occur when the stored energy, G, in that layer, of thickness h, equates to the oxide/metal interfacial fracture energy, G_c[8], i.e.

$$W^* h = G = G_c \tag{1}$$

W^* is the strain energy per unit volume of the oxide layer which, in the absence of mechanically imposed stresses, is the sum of the strain energy associated with oxide growth stresses and that due to differential strains during temperature changes[9]:

$$W^* = \left(\frac{\left(\sigma_g + E_{ox}(\Delta T)(\Delta \alpha) \right)^2 (1 - \nu_{ox})}{E_{ox}} \right) \tag{2}$$

where σ_g is the growth stress within the oxide layer, E_{ox} is the Young's modulus of the oxide layer, ν_{ox} its Poisson's ratio, ΔT the temperature change and $\Delta \alpha = (\alpha_m - \alpha_{ox})$ is the difference in linear expansion coefficients between metal and oxide. It is known that the growth stress for alumina can attain high values, of ~ -1GPa initially but often relaxes during continued exposure at high temperatures to values relatively small compared with thermal stresses induced during cooling to room temperature[10,11]. For present purposes, σ_g will be assumed to be small in relation to the thermal stresses, so that:

$$W^* \approx E_{ox}(1 - \nu_{ox})((\Delta T)(\Delta \alpha))^2 \tag{3}$$

From equations (1) and (3), G_c for oxide spallation is then obtained as:

$$G_c \approx E_{ox}(1 - \nu_{ox})((\Delta T)(\Delta \alpha))^2 h \tag{4}$$

If G_c is insensitive to oxide thickness, i.e. time at temperature, then for a fixed value of ΔT, the spallation event will occur at a critical oxide thickness, h_c, given as:

$$h_c = \left[\frac{G_c}{E_{ox}(1 - \nu_{ox})((\Delta T)(\Delta \alpha))^2} \right] \tag{5}$$

Alternatively, the critical temperature drop required to initiate spallation will be expected to vary as $h^{-\frac{1}{2}}$ so that, in this sense, spallation becomes easier as the oxide thickens. This functional dependence has been shown to apply to the spallation of a chromia layer from an austenitic steel[9,12] and from alumina on various Ni-based alloys[13]. This approach of defining the onset of spallation by a critical strain energy parameter is of general application and has been valuable in predicting the behaviour of steam-grown oxides in power plant also[14,15].

The extension of these concepts to TBC systems is readily achieved by the addition of the stored energy within the ceramic top coat to the total. Again, in this section, a flat thin-layer approximation will be used, as shown schematically in Figure 1, and strains will be generated through differential thermal contraction. Later in the paper, consideration will be given to out-of-plane stresses developed during the oxidation of non-planar bond coats.

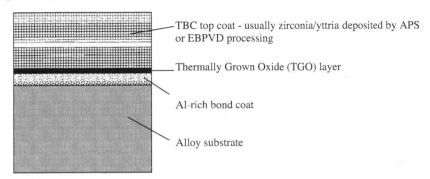

Figure 1. The notional TBC system used for calculation of the stored energy.

The total strain energy, G^{tot}, available to produce decohesion at the TGO/bond-coat interface is then given as:

$$G^{tot} = G_{tgo} + G_{ysz} \qquad (6)$$

where G_{tgo} is the strain energy within the TGO and G_{ysz} is that in the YSZ top coat. The strain energy in the oxide layer is similar to equation (4):

$$G_{tgo} = E_{tgo}\left(1 - v_{tgo}\right)\left(T - T_{ox}\right)^2 \left(\alpha_{sub} - \alpha_{tgo}\right)^2 h_{tgo} \qquad (7)$$

where T_{ox} is the oxidation temperature, T is the current temperature during the temperature change, α_{sub} is the linear expansion coefficient for the alloy substrate and α_{tgo} is its value for the TGO. The other parameters have been defined earlier and relate to the TGO. The strain energy, G_{ysz}, is given similarly as:

$$G_{ysz} = E_{ysz}\left(1 - v_{ysz}\right)\left(T - T_{dep}\right)^2 \left(\alpha_{sub} - \alpha_{ysz}\right)^2 h_{ysz} \qquad (8)$$

The parameters are as in equation (7) but now refer to the YSZ layer; T_{dep} is the temperature at which the top coat was deposited in an assumed stress-free condition.

Evaluation of G^{tot} can now be readily made using the values given in Table 1 together with T_{ox}=1373K, T_{dep}=1273K and T=373K. For typical values of h_{tgo}=3 μm and h_{ysz}=150 μm, G_{tgo}=72 J.m^{-2} and G_{ysz}=69 J.m^{-2} to give G^{tot}=141 J.m^{-2}. In principle, the quantity, G_{ysz} is available to produce decohesion at the top-coat/TGO interface and G^{tot} at the TGO/bond-coat interface and conceivably also for fracture within the TGO layer. In practice, failure has been found at both of the TGO interfaces.

Table I: Values of the Elastic Parameters Used in Evaluating G^{tot}.

	$E/10^9$, Pa	ν	$\alpha/10^{-6}$, K^{-1}
YSZ top coat	40	0.11	13.0
alumina TGO	380	0.24	7.9
alloy substrate	not needed	not needed	17.0

There is inevitable uncertainty in the evaluation of G_{ysz}, particularly for YSZ top coats deposited by the EBPVD process, because of anisotropy in the elastic constants and because the density of the coating is higher within approximately 20 µm of the TGO interface than elsewhere. Top coats deposited by APS processing have more uniform properties and the use of a single value of the Young's modulus is reasonable. The single estimated value for E_{ysz} will be adequate, however, for the present exercise which is intended to highlight the very large discrepancy between the energy stored within the YSZ and TGO layers and the intrinsic work of adhesion of the TGO/bond-coat interface.

First-principles calculations of the work of adhesion of alumina to β-NiAl have recently been undertaken by Carling and Carter[16]. Their calculated work of adhesion is the true interfacial fracture energy, G_c^{true}, and allows for the formation of new oxide and alloy interfaces and the removal of the original interface. The value they obtain for a clean original interface is only 0.66 J.m^{-2} which is over two orders of magnitude less than the total available strain energy in the TGO and YSZ top coat. This comparison shows that the TBC system is inherently mechanically unstable and should be expected to delaminate on the first cooling transient. It patently does not do so as can be appreciated from the compilation of failure times shown in Figure 2.

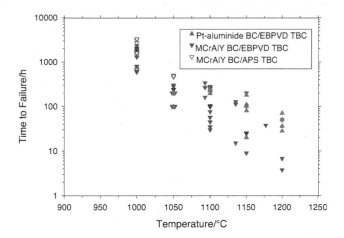

Figure 2. A comparison of TBC failure times[17].

The trend of decreasing lifetime with increasing temperature closely follows the temperature dependence of alumina thickening on the bond coat surface. Indeed, a lower bound line to the data approximates to the time required to form a 2.5-3.0 µm layer[17] indicating that there is an important role of oxidation in the failure process. However, within the scatter, there does not seem to be a direct correspondence between short lifetimes and thick oxides. Other factors are in play.

EFFECTIVE FRACTURE ENERGIES

It has been known for many years[18-20] that the spallation of the oxide layer from conventional alloys is not determined by the true work of adhesion of the oxide/metal interface but rather by an effective fracture energy. This can be 1-2 orders of magnitude greater than the true value but is the quantity that needs to be used in the energy balance of equation (1). These large effective fracture energies indicate that a substantial (>90%) fraction of the stored energy within the oxide layer is dissipated by deformation and is not directly used in the fracture (spallation) process. Useful insights have been obtained from finite element (FE) computations[18,19] of the kinetics of interfacial crack growth during cooling. These show that considerable creep relaxation (principally in the alloy) of interfacial crack-tip stresses can occur during cooling and that this inhibits the growth of the crack during the early stages of the cooling transient. Crack growth then takes place when the current temperature has reduced sufficiently that creep relaxation rates are small. An example of the predicted crack growth kinetics for a 5-μm thick alumina layer on a Haynes 214 alloy during cooling from 1100°C (1373K) is shown in Figure 3. The region of zero crack growth rate is clear for both the cooling rates used but crack growth commences at a higher current temperature at the higher cooling rate. This reflects the reduced ability of creep relaxation to take place at higher cooling rates. The dotted line in this figure reflects the growth kinetics expected from the critical strain energy criterion when the stored energy is released by crack growth at a critical temperature drop ΔT_c. It can be appreciated that the predicted kinetics, *in the presence of creep relaxation*, approximate well to this ideal. For the higher strain rate, $|\Delta T_c| = 397$°C and for the lower strain rate, $|\Delta T_c| = 542$°C. These values can be used in equation (4) together with the parameter values given in Table 1 to give corresponding values of G_c of 23 J.m^{-2} and 42 J.m^{-2}, for the higher and lower cooling rates, respectively. These values are appreciably higher than the true work of adhesion and show the importance of creep relaxation in determining the growth kinetics of the interfacial cracks. They also show that G_c has no unique value but will vary with test conditions, in this example the cooling rate.

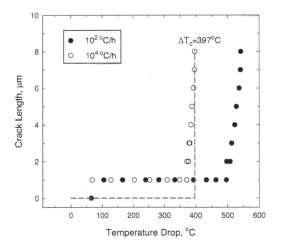

Figure 3. Finite element computations of the kinetics of growth of a wedge crack along the alumina/Haynes 214 alloy interface during cooling from 1100°C at two cooling rates.

Similar FE computations have been reported[6] for wedge crack growth along the TGO/bond-coat in a TBC system consisting of a Haynes 230 substrate alloy, a NiCrAlY bond coat and APS YSZ top coat. These results again show a period of zero crack growth during the early stages of cooling where bond coat creep relaxation reduces crack-tip stresses.

The above discussion demonstrates that stress relaxation and associated crack-tip blunting by local creep processes delays the failure by spallation of TBC systems and leads to high values of effective fracture energies. It is also feasible, however, for creep and plasticity to facilitate the fracture process. This is the basic premise behind mechanisms such as ratcheting[21,22] and rumpling[23-27] where deformation of both bond coat and TGO create, in the former case, void-like damage in the vicinity of the TGO/bond-coat interface and, in the latter case, increased roughening of the bond coat and associated out-of-plane stresses. In both cases, the driving force for the change in interface morphology is the relaxation of compressive stresses within the TGO by its lengthening. These stresses can be induced during oxide growth or as a result of thermal cycling. There is experimental evidence for the ratcheting mechanism[21] although the incidence of failure by this route does not seem widespread. Rumpling, i.e. roughening, of the TGO/bond-coat interface has been observed under thermal cycling but *only* in the absence of the ceramic top coat and does not seem to occur in standard TBC systems.

The difficulty with both these mechanisms is the mechanical constraint imposed by the YSZ top coat. For the rumpling process, for example, it is necessary for the bond coat to be sufficiently weak to allow relatively large-scale deformation but it must also be strong enough to induce similar local displacement within the top coat. These two contradictory requirements of the bond coat creep strength may be satisfied only within a limited temperature and composition range, if at all. Similarly, the ratcheting process and the creation of voids within the bond coat occurs at constant volume and the space created during void formation must be accommodated somewhere. An obvious route is for the displaced bond coat matter to produce an upward displacement of the top coat in much the manner envisaged for rumpling. However, there remains the contradictory requirement of low creep strength to produce the morphological changes and the need of high creep strength to displace the top coat and allow the changes to occur.

STRESSES ASSOCIATED WITH THE GROWING OXIDE LAYER

The formation of the alumina TGO occurs in a mechanically constrained environment between the bond coat and YSZ top coat. Conversion of aluminium into its oxide results in a volumetric expansion and an associated strain of ~25% most of which is directed perpendicular to the bond coat surface[7]. For flat interfaces, as have been considered so far, this expansion would simply displace the top coat outwards and no stresses would arise. Real bond coat surfaces, of course, are not flat and the extent of their roughness will have a significant influence on stress development within the system.

This can be understood from the schematic diagram of Figure 4 which considers local out-of-plane displacements around a bond coat asperity. Stresses will develop in an out-of-plane sense within the top coat whenever there is a variation of upward displacement rates along the bond coat surface. These may arise from differences in oxidation growth rates but can also arise simply from geometrical effects even when the intrinsic oxidation rates are everywhere constant. These strains may lead to cracking (termed continuity cracking) at the oxidation temperature under isothermal testing conditions. The direction of TGO growth (Figure 4) is perpendicular to the local bond coat surface so that at peaks and troughs or on a planar surface the direction of growth is wholly vertical. However, at any point away from these the growth direction will have a vertical and horizontal component and the former will have a magnitude less than that at the peaks and troughs. Out-of-plane strains and associated tensile stresses will then be imposed upon the ceramic top coat if continuity is to be maintained. These stresses will be expected to occur along the flanks of the bond coat asperities.

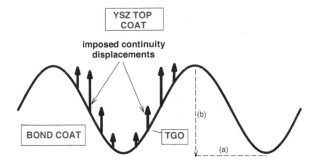

Figure 4. Continuity strains are imposed upon the YSZ top coat at the oxidation temperature to accommodate non-uniform upward displacements resulting from TGO growth on a non-planar TGO/bond-coat interface.

This expectation of out-of-plane tensile stress development at the oxidation temperature is borne out in recent FE computations[7,28]. In this work, creep, plastic and elastic mechanical properties were used and the alumina TGO was modelled to thicken during the simulated computer exposure. This simulated oxidation not only converted near-surface mesh elements to produce new oxide but also the associated volume expansion was incorporated. The computed distribution of out-of-plane stresses after 100 hours exposure at 1100°C, when the oxide had thickened to 3.3 μm, is shown in Figure 5 for two values of bond coat roughness. This roughness is defined as the ratio (b/a) shown in Figure 4.

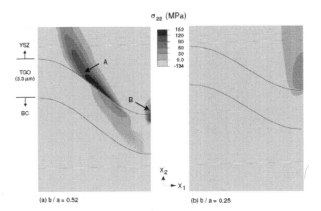

Figure 5. Predicted out-of-plane stresses at 1100°C for two roughness levels after 100 hours exposure. Maximum stresses, resulting from TGO growth, develop in the YSZ top coat[7].

These computations were undertaken for a Pt-aluminide bond coat and an EBPVD YSZ top coat. Full details of the materials' properties and computational approach can be found in Busso et al.[7]. The results show that the tensile stresses that develop at the base of the top coat, along the flanks of the bond coat asperity, at the oxidation temperature are strongly dependent on surface roughness. Halving the (b/a) ratio from 0.52 to 0.25 reduces these tensile stresses from ~150 MPa to essentially zero. Cooling to 25°C results in the predicted stress distribution shown in Figure 6 (note the different stress scale from Figure 5). For the rougher bond coat, the flank stresses within the top coat are unaffected by this cooling transient but very large out-of-plane tensile stresses of ~700 MPa have now developed within the valley regions of the top coat. Tensile stresses of similar magnitude also act across the YSZ/TGO interface at this location and of ~400 MPa across the TGO/bond-coat interface. Again, the less-rough surface is much less stressed.

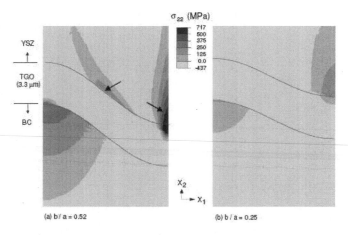

Figure 6. Predicted out-of-plane stresses at 25°C for two roughness levels after 100 hours at 1100°C.

This combination of out-of-plane tensile stresses developed at the oxidation temperature along asperity flanks and those produced in the valley regions during cooling is particularly onerous. For rough regions of bond coats it provides the opportunity for flank cracks to form at the oxidation temperature (or lower) and to propagate across the valley regions during cooling. Examples of this type of cracking are shown in Figure 7 for an EBPVD-YSZ/Pt-aluminide TBC system after 2 hours exposure at 1200°C. The white arrows in this figure indicate locations where cracks have traversed valley locations and, in some instances, have penetrated the TGO. At these locations, the local bond coat roughness is particularly high with (b/a) values in the range 0.6 to 1.0. It is expected that local out-of-plane stresses even higher than those predicted above would develop at these regions.

Whether or not these cracks can develop into a longer delamination that might permit final failure to occur by buckling, for example, depends on the consistency of bond coat surface roughness. This is because the strain energy associated with the continuity strains is localised. Any cracks that form will not have sufficient driving force to propagate unless adjacent regions of the bond coat have a similar roughness. Implicit in this statement is the recognition that the large values of energy stored within the top coat during cooling do not contribute significantly to the formation of these microcracks. This is probably reasonable since no mechanism seems available and it is generally accepted that the release of the top coat energy is the driving force for final stage buckling of these coatings.

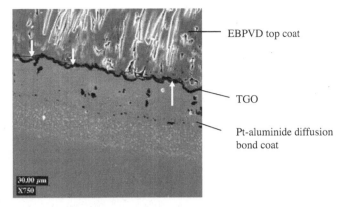

Figure 7. Microcracks (arrowed) formed in an EBPVD-YSZ/Pt-aluminide TBC system after an exposure of 2 hours at 1200°C.

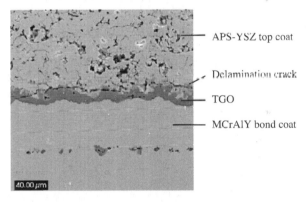

Figure 8. Interlinked microcracks to form a delamination zone at the base of the top coat in an APS-YSZ/MCrAlY TBC system after 400 hours isothermal exposure in air at 1100°C[8].

The TBC system shown in Figure 7 uses an EBPVD top coat and has an unusually rough bond coat surface for this deposition process. EBPVD top coats adhere well to much flatter surfaces and the usual delamination path is at the TGO/bond-coat interface. It is presumed that this is because no mechanisms exist to produce the top coat cracks shown in Figure 7 on flatter surfaces (Figures 5 and 6). APS top coats, on the other hand, are deposited intentionally on rough bond coat surfaces and the delamination path is usually within the top coat with some penetration into the oxide. It is likely that this occurs by the mechanisms discussed in this section. An example of such a system where a large delamination crack has developed, but final spallation has not occurred, is shown in Figure 8.

The present discussion has shown that the formation of cracks at the base of the top coat is reasonably understood. By contrast, the mechanisms for crack formation at the TGO/bond-coat interface remain unclear. Further discussion on this is beyond the scope of the present paper but it is an important area since this interface appears to be favoured for sensibly flat interfaces. Various mechanisms have been proposed in the literature and some have been briefly mentioned here. These others include void formation at the TGO/bond-coat interface and the segregation of harmful elements such as titanium and sulphur. The present author favours the growth of wedge cracks along the (flat) TGO/bond-coat interface since their growth during cooling can be driven by the substantial strain energy that develops within the oxide layer. It is entirely feasible that more than one mechanism operates, however, and the subject is still a matter of debate.

Nevertheless, the importance of bond coat surface roughness is mechanistically established. It has also been confirmed experimentally in elegant work by Yanar et al.[29] in which the bond coat roughness was systematically changed in a TBC system using an EBPVD YSZ top coat. They found that lifetimes increased as surface roughness decreased by about an order of magnitude over the roughness range studied. Interestingly, this is the typical scatter found within and between production TBC batches, as can be appreciated from the data in Figure 2, and it is certainly tempting to ascribe much of this scatter to poorly controlled variation in the bond coat surface roughness.

BREAKAWAY OXIDATION

Chemical Failure

The discussion in the preceding section has focussed on the influence of the growth of an alumina layer of uniform thickness on the development of out-of-plane stresses in the vicinity of the TGO interfaces. The formation of this oxide selectively removes aluminium from the bond coat, reducing its concentration adjacent to the oxide layer. This local depletion can become so severe that the protective alumina layer can no longer be maintained; this process is termed "chemical failure". Two generic routes to chemical failure have been identified[30]. At extreme depletion, where Al levels have dropped close to zero, it becomes thermodynamically possible for other alloy constituents, e.g. Cr, to decompose alumina to form their own oxide or to form the oxide underneath the alumina layer if oxygen continues to diffuse through this. This route is termed "Intrinsic Chemical Failure" (InCF) and has been shown to occur in alumina-forming ferritic alloys[31] and a NiCrAlY overlay coating[32]. The second route to chemical failure can occur at moderate depletion levels but where the local concentration of aluminium is insufficient to re-form a protective layer. Typical concentration levels where this could occur would be $\leq$ 2 wt.% Al. This process is termed "Mechanically Induced Chemical Failure" (MICF) and would be triggered by mechanical damage to the protective layer. It is the more likely process to initiate chemical failure at bond coats within TBC systems simply because stresses are present and the rough surfaces of many bond coats can restrict re-supply of aluminium to the surface regions.

The influence of surface topology on local aluminium depletion can be appreciated from Figure 9 which shows 2-dimensional finite difference computations[33] for an asperity on the surface of a NiCrAlY coating. The computations were undertaken for a notional test temperature of 1100°C and allowed for the continued growth of a protective alumina layer around the asperity and elsewhere on the coating surface. Ternary diffusion coefficients were used. It is clear that enhanced depletion of aluminium occurred within the asperity and this was a result of the increased rate of removal to the oxide layer (a larger surface area) and a restricted rate of supply through the neck region. As a consequence, mechanical failure of the oxide around the asperity would lead to the initiation of MICF since nowhere within the asperity is there sufficient aluminium to re-heal the alumina layer. A consequence would be the gradual consumption of the asperity through the formation of fast-growing Ni (and Co, where applicable)-rich oxides. This process has been shown to occur in a TBC system with a MCrAlY bond coat[33] but a particularly nice example from earlier work by Hsueh et al.[34] is reproduced in modified form in Figure 10.

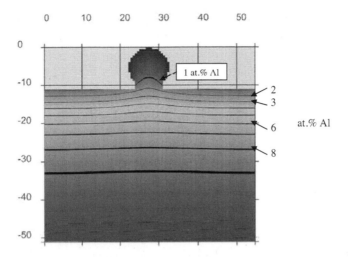

Figure 9. Finite difference predictions of aluminium concentration in and around a 2-dimensional asperity on a 100 μm thick MCrAlY coating after 1h oxidation in air at 1100°C. The solid lines show the Al concentration contours. The dimensions of the model are shown in microns[33].

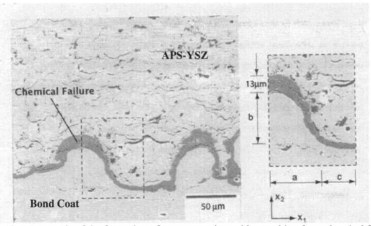

Figure 10: An example of the formation of non-protective oxides resulting from chemical failure atbond coat asperities in a MCrAlY/APS-YSZ TBC system[34].

Effect on Stress Development

The formation of fast-growing oxides at asperity tips is expected to have a profound effect on stress development within the TBC system. To assess this, the morphology of bond coat and variation in oxide thickness shown in Figure 10 has recently[35] been modelled by finite element methods. As described earlier, the oxide was allowed to grow as part of the computational process and the associated volumetric strains were also incorporated. The computations were undertaken for isothermal oxidation at 1100°C for a total exposure of 200 hours. A protective alumina layer grew during the first 100 hours but then chemical failure was initiated. The oxide thickness distribution at the end of the 200 hour period reproduced the boxed area in Figure 10. Full details can be found in Busso et al.[35].

The predictions for the out-of-plane stresses developed at the oxidation temperature after 200 hours exposure are shown by the contour plot of Figure 11 for two surface roughnesses. The expected development of tensile stresses at the base of the top coat adjacent to the flank of the bond coat asperity is clear. It can also be seen that reducing the bond coat surface roughness again has a beneficial effect, reducing peak stresses from ~ 650 MPa to ~ 200 MPa for a change in (b/a) values (see Figure 10) from 1.0 to 0.25. These are remarkably high stresses to be developed at 1100°C and would be expected to prejudice the integrity of the system, particularly for the rougher surface. By contrast, the normal stress across the YSZ/TGO interface at the apex of the asperity was found to be highly compressive ~ -900MPa for the rougher surface and to be close to zero elsewhere around the asperity. Interfacial damage at temperature is unlikely.

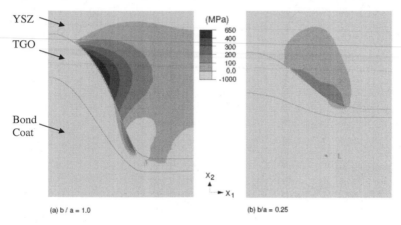

Figure 11. Contour plot of the out-of-plane, σ_{22}, stress which develops at 1100°C after 200 hours oxidation in air at that temperature for (a) a b/a ratio of 1.0 and (b) a ratio of 0.25[35].

Cooling this structure to 25°C after 200 hours exposure results in a substantial increase to > 900 MPa in the peak out-of-plane tensile stresses at the base of the top coat as shown in Figure 12. In contrast to the situation at temperature, significant tensile stresses, around 400 MPa, had now developed at the TGO/bond-coat interface at the apex of the asperity.

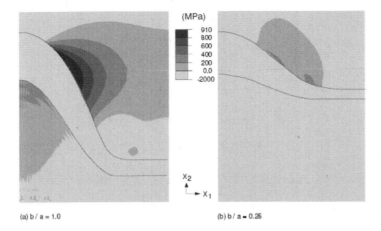

(a) b / a = 1.0   (b) b / a = 0.25

Figure 12. Contour plots of the out-of-plane, σ_{22}, stress at 25C after 200 hours oxidation at 1100°C for bond coat roughness parameters of (a) b/a = 1.0 and (b) b/a = 0.25[35].

CONCLUDING SUMMARY

It has been shown that typical TBC systems during cooling contain stored strain energy within the YSZ top coat and alumina TGO some 1-2 orders of magnitude larger than the true work of adhesion of the TGO/bond-coat interface. These TBC systems are inherently mechanically unstable but patently do not delaminate and useable lifetimes ranging from ~1000 hours at 1000°C to ~10 hours at 1200°C are readily achieved. Their longevity arises because creep processes are active, particularly within the bond coat. Creep relaxation and blunting of crack tips at the TGO/bond-coat interface dissipates strain energy during cooling, for example. Creep processes, as such, are also needed to produce crack-like damage in some postulated mechanisms. The overall effect is that fracture damage increases over time and its development is associated with energy dissipation. Energy balance approaches to failure can be applicable, provided an effective fracture energy is used. No unique value exists for this, however, and variation will arise with cooling rate and exposure temperature. This type of approach has been used over many years for the prediction of protective oxides from conventional alloys.

Bond coat surface roughness has an important influence on TBC endurance. The growth of the oxide layer on a non-planar interface generates tensile continuity strains within the top coat adjacent to the flanks of bond coat asperities. Finite element computations show that out-of-plane tensile stresses at the base of the top coat resulting from the growth (and associated volume expansion) of a uniform alumina layer can exceed 150 MPa at the oxidation temperature of 1100°C after 100 hours exposure. These stresses arise for a moderately rough bond coat surface having a (b/a) ratio (defined in Figure 4 of the text) of 0.52. Cooling this structure retains these continuity stresses but also develops high (~700 MPa) out-of-plane stresses in the top coat over valley regions. It is suggested that this combination of stresses, developed at temperature and during cooling, accounts for the formation of lateral microcracks in the top coat between bond coat asperities. Having a less-rough interface, with (b/a) = 0.25, essentially eliminates such large stresses.

An additional degradation process of importance, particularly in MCrAlY bond coats, is chemical failure at aluminium-depleted asperities. As a result, fast-growing non-alumina oxides can form and develop substantial out-of-plane tensile strains within the top coat, again along the flanks of the asperity. Finite element analyses indicate that out-of-plane tensile stresses as high as 650 MPa can then form for rough surfaces, (b/a) = 1.0, and ~100 MPa for (b/a) = 0.25. These stresses result from the expansion strains associated with oxide growth and will be present at the oxidation temperature of 1100°C. Cooling these structures to 25°C produces an increase in these stresses to ~900 and ~200 MPa, respectively. In addition, cooling was found to have produced tensile tractions across the TGO/bond-coat interface of ~400 MPa for the rougher surfaces.

This combination of results emphasises the importance of producing as flat a bond coat surface as feasible. The dependence of the various delamination mechanisms on surface roughness are only now being assessed quantitatively using realistic numerical models which allow for creep processes and the volume expansion due to oxide formation. Work in this area is incomplete but some rather qualitative trend lines are shown schematically in Figure 13 which relates TBC lifetimes to the bondcoat surface roughness. This figure should be read within the context of the text and the suggested order of magnitude (b/a) values at the boundaries between the three domains are speculative but seem consistent with the numerical results and observations.

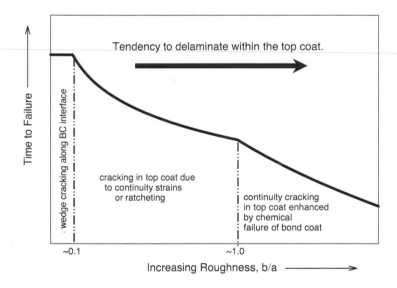

Figure 13. A schematic chart showing qualitatively the influence of bond coat surface roughness on oxidation-induced mechanical failure processes in TBC systems. Further details are provided in the text.

REFERENCES

1. D.J. Wortman, B.A. Nagaraj and E.C. Dunderstadt, *Mater. Sci. Eng.*, **A121**, 433, (1989).
2. J.T. Demasi-Marcin, K.D. Scheffler and S. Bose, *J. Eng. Gas Turbines Power*, **112**, 521, (1990).
3. A.H. Bartlett and R.D. Maschio, *J. Am. Ceram. Soc.*, **78**, 1018, (1995).

4. J.A. Haynes, E.D. Rigney, M.K. Ferber and W.D. Porter, *Surf. Coat. Tech.*, **86-87**, 102, (1996).
5. M.P.Taylor, H.E. Evans, S. Gray and J.R. Nicholls, *Mater. Corr.*, in press.
6. H.E. Evans and M.P. Taylor, *J. Corr. Sci. Eng.*, **6**, H011, (2003).
7. E.P. Busso, Z.Q. Qian, M.P. Taylor and H.E. Evans, *Acta Mater.*, **57**, 2349, (2009).
8. U.R. Evans, "An Introduction to Metallic Corrosion", p. 194, Edward Arnold, London, (1948).
9. H.E. Evans and R.C. Lobb, *Corr. Sci.*, **24**, 209, (1984).
10. D.M. Lipkin, D.R. Clarke, M. Hollatz, M. Bobeth and W. Pompe, W., *Corr. Sci.*, 39, 231, (1997).
11. H. Echsler, E.A. Martinez, L. Singheiser and W.J. Quadakkers, *Mater. Sci. Eng.*, **A384**, 1, (2004).
12. H.E. Evans, *Mater. Sci. Tech.*, **4**, (1988), 415.
13. K. Bouhanek, D. Oquab and B. Pieraggi, *Mater. Sci. Forum*, **251-254**, 33, (1997).
14. J. Armitt, D.R. Holmes, M.I. Manning, D.B. Meadowcroft and E. Metcalfe, EPRI Report FP-686, (1978).
15. I.G. Wright, M. Schütze, S.R. Paterson, P.F. Tortorelli and R.B. Dooley, EPRI International Conference on Boiler Tubes and HRSG Tube Failures and Inspections, San Diego, (2004).
16. K.M. Carling and E.A. Carter, *Acta Mater.*, **55**, 2791, (2007).
17. R.D. Jackson, M.P. Taylor and H.E. Evans, in *Microscopy of Oxidation 7*. Eds. G.J. Tatlock and H.E. Evans, Science Reviews, St. Albans, UK.
18. H.E. Evans, G.P. Mitchell, R.C. Lobb and D.R.J. Owen, *Proc. Roy. Soc. A*, **440**, 1, (1993).
19. H.E. Evans, A. Strawbridge, R.A. Carolan and C.B. Ponton; *Mater. Sci. Eng.*, **A225**, 1, (1997).
20. A. Galerie, F. Toscan, E. N'Dah, K. Przybylski, Y. Wouters and M. Dupeux, *Mater. Sci. Forum*, **461-464**, 631, (2004).
21. D.R. Mumm, A.G. Evans and I.T. Spitsberg, *Acta Mater.*, **49**, 2329, (2001).
22. A.M. Karlsson, J.W. Hutchinson and A.G. Evans, *Mater. Sci. Eng.*, **A351**, 244, (2003).
23. P. Deb, D.H. Boone and T.F. Manley, *J. Vac. Sci. Technol. A*, **5**, 3366, (1987).
24. Z. Suo, *J. Mech. Phys. Solids*, **43**, 829, (1995).
25. M.Y. He, A.G. Evans and J.W. Hutchinson, *Acta Mater.*, **48**, 2593, (2000).
26. V.K. Tolpygo and D.R. Clarke, *Acta Mater.*, **48**, 3283, (2000).
27. V.K. Tolpygo and D.R. Clarke, *Acta Mater.*, **52**, 5129, (2004).
28. E.P. Busso, L. Wright, H.E. Evans, L.N. McCartney, S.R.J. Saunders, S. Osgerby and J. Nunn, *Acta Mater.*, **55**, 1491, (2007).
29. N.M. Yanar, F.S. Pettit and G.H. Meier, *Metall. Mater. Trans.*, **37A**, 1563, (2006).
30. H.E. Evans, A.T. Donaldson and T. C. Gilmour, *Ox. Met.*, **52**, 317, (1999).
31. G. Strehl, D. Naumenko, H. Al-Badairy, L.M. Rodrigeuz Lobo, G. Borchardt, G.J. Tatlock and W.J. Quadakkers, *Mater. High Temp.*, **17**, 87, (2000).
32. P. Niranatlumpong, C.B. Ponton and H.E. Evans, *Ox. Met.*, **53**, 241, (2000).
33. M.P. Taylor, W.M. Pragnell and H.E. Evans, *Mater. Corr.*, **59**, 508, (2008).
34. C-H. Hsueh, J.A. Haynes, M.J. Lance, P.F. Becher, M.K. Ferber, E.R. Fuller Jr., S.A. Langer, W.C. Carter and W.R. Cannon, *J. Amer. Ceram. Soc.*, **82**, 1073 (1999).
35. E.P. Busso, H.E. Evans, Z.Q. Qian and M.P. Taylor, *Acta Mater.*, **58**, 1242, (2010).

STRENGTH DEGRADING MECHANISMS IN PLASMA SPRAY COATED SILICON NITRIDE

Ramakrishna T. Bhatt
Vehicle Technology Center
US Army Research Laboratory
Glenn Research Center at Lewis Field
21000 Brookpark Rd.
Cleveland, OH 44135

Dennis S. Fox
NASA Glenn Research Center
21000 Brookpark Road
Cleveland, Ohio 44135

Abdul-Aziz Ali
Cleveland State University
2121 Euclid Avenue
Cleveland, OH 44115

ABSTRACT

Two high temperature grades of monolithic silicon nitrides were coated with a plasma sprayed barium aluminum strontium silicate (BSAS) based environmental barrier coating (EBC). Their room temperature flexural strengths were then measured. The EBC coated specimens showed nearly 50% loss in strength. Various factors such as substrate preparation methods, plasma spray damage to the substrate, thermal residual stresses, and stress raisers influenced the strength. To determine the strength degrading mechanism, the role played by various phases of the coating process on the substrate strength has been examined, and maximum thermal residual stresses generated in the coating has been modeled. Results indicate that three different types of flaws related to the coating deposition process in combination with tensile residual stress cause strength degradation. Of these flaws, two can be eliminated by pre and post processing treatments, but well bonded splats cannot be avoided. The splats and preexisting flaws on the substrate surface act as a stress raiser due to imposed tensile thermal residual stress of the coating, and thus cause strength degradation.

INTRODUCTION

Monolithic silicon nitride materials offer a variety of performance advantages over the best metallic alloys for high-temperature components in advanced engines for power and propulsion. These advantages are primarily based on higher temperature capability (~ 200 to 300^0C more than the current metal design), and lower density (~30-50% of the metal density). For the last three decades, several grades of silicon nitride ceramics have been developed for high temperature components such as stators, rotors, scrolls, shroud, blades and nozzle vanes [1, 2]. Of these grades, two high temperature silicon nitrides have adequate properties for use as first stage turbine components where temperature can reach as high as 1500^0C. Although silicon nitride displays excellent high temperature properties, high conductivity and oxidation resistance in static air, it undergoes surface recession due to simultaneous formation and volatilization of silica similar to all other silicon based ceramics and SiC/SiC composites when exposed to a combustion environment containing moisture at temperatures> 1100^0C [3, 4, 5]. To protect the SiC/SiC composites from surface recession, environmental barrier coatings (EBC) have been developed. One key example is a multilayered coating having a barium strontium aluminum silicate (BSAS) and rare earth silicate top coat [6, 7]. The BSAS and rare earth silicate based

EBCs have upper temperature capabilities of ~1316^{0}C [7] and ~1482^{0}C [8], respectively for applications over thousands of hours. The EBC in general consists of 2 or more layers, and each of these layers serve specific purpose. The layer on top of the substrate is the bond coat layer which prevents diffusion or migration of sintering additives and impurities from the substrate into the top coat. The layer on top of the bond coat is an intermediate layer which prevents reaction between the top and bond coats. The layer on top of the intermediate layer is the top coat which prevents diffusion of moisture into the substrate. The thickness of the individual layers can vary from 75 μm to125 μm depending on the type and intended maintenance schedule of the components. These multilayered coating can be applied on the substrate by plasma spray (PS), electron beam physical vapor deposition (EBPVD), or slurry coating. Because of the differences in elastic and fracture properties, and thermal expansion between different layers of the EBC and the substrate, the EBC can have an adverse or beneficial effect on the mechanical properties of the substrate. Large difference in thermal expansion between the substrate and the EBC causes thermal residual stresses, which can change after long term exposure to high temperatures in oxidizing or combustion environment. In addition, the EBC can also act like a notch, affecting the strength of the substrate. Studies have shown that in-plane properties of SiC/SiC composites are not affected by EBC deposition by plasma spray, but similar coating on monolithic silicon nitride causes up to 50% strength loss [9-12] which is well below the design strength. Such a large loss in strength is not acceptable for designing this material for any high temperature applications. A variety of factors such as tensile residual stress in the coating, possible damage created by the plasma spray process on the substrate surface, or lack of crack deflecting mechanism at the substrate/ coating interface may contribute to strength degradation. However, the dominant factor or factors responsible for strength degradation have not been fully understood. Coatings in general do not degrade strength of monolithic ceramics. For example, thermally grown rare earth silicate coating or thin (<1 μm) chemically vapor deposited mullite coating on silicon nitride or plasma sprayed mullite coating on silicon carbide tubes had no effect on the strength of the substrate [13-16].

The objective of this study is to determine the strength degradation mechanism(s) of silicon nitride coated with BSAS based EBC by air plasma and to develop finite element models to predict thermal residual stresses in the coating and conditions of crack deflection and propagation from the coating into the substrate

EXPERIMENTAL PROCEDURE

Two grades of monolithic silicon nitride billets (AS-800 and SN282) of dimensions ~100mm (L) x100mm (W) x 3.2mm (T) were purchased from commercial vendors. The Honeywell Engine Components Division, Torrance, CA and Kyocera Industrial Ceramics Corporation, Vancouver, WA provided AS-800 and SN282 grades, respectively. The billets were ground and cut into bend bars of dimensions 45mm (L) x 4mm (W) x 3mm (T) using diamond impregnated metal wheels and blades according to ASTM standard procedure # C 1161-02C. The long edges of the test specimens were chamfered to minimize failures initiated by edge chips. In general, the protocol for plasma spray coating consists of surface preparation of the substrate, depositing the coating on the heated substrate and then heat treating the substrate. The surface preparation is needed for better adherence of the coating and is typically accomplished by grit blasting, molten salt etching, or thermal treatment. To understand the flaws created in different phases of plasma spray coating, the coating(s) was or were deposited on as-machined as well as surface prepared bend bars. Surface preparation methods included grit blasting and molten salt etching. On some bend bars, only one layer of silicon or mullite was coated, but on others a layer of silicon was deposited first, followed by a mixture of mullite and barium aluminum silicate (BSAS), and then a layer of BSAS. The thickness of individual coating was either 75 μm or 125 μm. The coatings were applied by air plasma using a procedure described in reference 17. The PS coatings were applied only on one side of the specimen.

TESTING AND CHARACTERIZATION METHODS

Four-point flexural strength tests were performed at a cross-head speed of 0.51 mm/min with inner and outer spans of 20 and 40 mm, respectively. Room temperature tests were conducted in air using SiC fixtures, following ASTM standard test procedures ASTM C 1161-B. At each test conditions three specimens were tested. Fracture surfaces of the failed specimens were examined in a scanning electron microscope.

RESULTS AND DISCUSSION

Figure 1 shows room temperature flexural strengths of uncoated and environmental barrier coated silicon nitride specimens (AS-800 and SN-282 grades). The specimens were grit blasted before PS deposition. The multilayered EBC consists of a layer of silicon bond coat followed by a mixture of mullite and BSAS as intermediate coat, and then a layer of BSAS top coat. The data shows that irrespective of the grade of the silicon nitride, ~50% loss in strength of the substrate was noticed after coating, which is consistent with the results of several studies [10-12]. To determine coating thickness effect on strength, the same two silicon nitrides were coated with silicon or mullite layer up to a thickness of 125 μm without any surface pretreatment or post thermal treatment. The variation of room temperature flexural strength for two grades of silicon nitride with thickness is shown in figure 2.

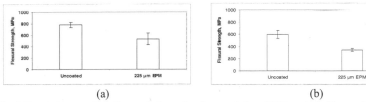

(a)   (b)

Fig. 1 Room temperature flexural strength of uncoated and EB coated silicon nitride:
(a) AS-800; (b) SN-282.

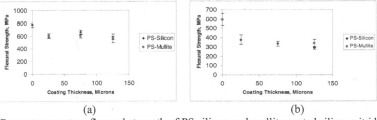

(a)   (b)

Fig. 2 Room temperature flexural strength of PS silicon and mullite coated silicon nitride showing the influence of coating thickness on strength: (a) AS-800; (b) SN-282.

Results indicate weak influence of coating thickness on strength. Since plasma spray coating involves several steps, it is difficult to decipher which step leads to strength degradation in figures 1 or 2. To investigate further, the strength of the substrates after each step of the processing step was measured. In some batches, the uncoated specimens were exposed to plasma flame alone and in other batches the uncoated specimens were grit blasted and molten salt etched. Then PS silicon coating was deposited on both as-machined and grit blasted surfaces of the specimens. Figure 3 shows the strength data for specimens subjected to various stages of plasma spray processing along with strength data of uncoated, as-machined specimens. The figure indicates that plasma flame exposure alone does not degrade strength of the silicon nitride, but grit blasting or molten salt etching or deposition PS silicon coating on as-machined as well as grit blasted surfaces does degrade the strength. Although not shown, a grit blasted specimen coated with a multilayered coating also showed strength degradation similar to as-machined specimen coated with a multilayered EBC. This indicates that the plasma sprayed coating itself is causing strength degradation irrespective of the surface preparation method. Loss in strength can be due to tensile residual stress in the coating, coating process damaging the substrate, or by a crack formed in the coating traveling into the substrate due a lack of a crack deflection mechanism at the coating/ substrate interface. In this paper the first two factors were evaluated in detail and the last factor is being investigated.

To determine the critical flaw responsible for failure of the uncoated and EB coated silicon nitride specimen, fractography was conducted using a scanning electron microscope (SEM). Based on nearly 100 fractographs, failure origins were classified as surface/interface, volume, edge, and coating flaws. The surface flaws include pits or depressions created by the surface modification methods and large beta grain on the surface or at an angle. The volume flaws include un-sintered regions and voids. The coating flaws include blobs of coating or a well bonded regions of the coating on the surface of the substrate which possibly acts as stress riser. The size of these flaws varies between 100 and 150 µm. Some examples of these flaws are shown in figures 4, and 5. The strength determining flaws in uncoated and coated SN-282 silicon nitride bars subjected to pre and post plasma coating are complied in Table I. The table shows that failure origins for coated silicon nitride vary depending on different phases of the coating process. The surface preparation methods used in this study damaged the substrate surface even before the plasma spray and all coated substrates showed edge effects. Based on Table I the following conclusions can be made. First, uncoated silicon nitride bars without any surface treatment fail from a volume or surface flaw as expected. Second, uncoated as well as coated silicon nitride bars subjected to surface modification treatments alone such as grit blasting and molten salt etching fail predominantly from the resultant surface pits. Third, the preponderance of failures in coated bars with or without any pre-surface treatment occur from corner or edge flaws. Fourth, the edge or corner failures in the coated bars can be reduced or eliminated by chamfering. Five, when surface and edge flaws are suppressed, the coated bars fail from a flaw within the coating or a volume flaw within the substrate.

In general, to retain greater fraction of as-produced strength, the coating process should not create a flaw larger than that present in as-produced substrate. The other factor that also influences the strength of the coated substrate is the nature and magnitude of the thermal residual stresses in the coating. If thermal residual stress is compressive, then the coated substrate may show higher strength than uncoated substrate. On the other hand, if the thermal residual stress is tensile, the coated substrate may show lower strength.

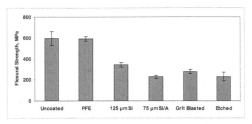

Fig. 3 Room temperature flexural strength of uncoated, plasma flame exposed, PS silicon coated, grit blasted, and molten salt etched SN-282 grade silicon nitride showing the influence of different stages of coating process on strength. The 125 μm silicon coating was deposited on the as-machined surface, and 75 μm coating deposited on as machined surface and then annealed at 1300°C for 20hrs.

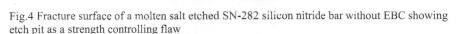

Fig.4 Fracture surface of a molten salt etched SN-282 silicon nitride bar without EBC showing etch pit as a strength controlling flaw

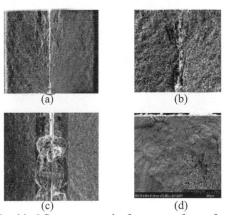

Fig.5 Typical strength critical flaws seen on the fracture surfaces of coated silicon nitride: (a) edge flaw, (b) large β Si₃N₄ grain at the interface, (c) coating flaw (splat) and (d) volume flaw (un-sintered region). In the figure both halves of the fracture surfaces are combined to identify the flaw.

FINITE ELEMENT MODELING

To determine the magnitude of residual stresses developed during deposition of EBC on silicon nitride, a finite element model (FEM) was developed to calculate in-plane, (x and y directions) and through-the-thickness (z direction) residual stresses. The physical and thermal properties of stand alone layers of EBC, i.e. PS Silicon, PS mullite+ BSAS and PS BSAS required for the model were obtained from reference 15. Results of the model (Fig. 6) indicate that z-direction stresses are small and negligible, but maximum in-plane stresses can be significant for single layer of silicon or mullite, or multilayered EBC. For the calculation, the substrate and the different layers of the coating are assumed to be perfectly elastic and bonded during and after deposition. In practice, during plasma deposition process, the coating material is partially molten and the stresses in the coating should be lower than those predicted by the model. Based on the modeling results it appears that tensile residual stress can be a factor influencing strength of PS coated silicon nitrides.

Table I   Strength critical flaws and failure statistics for uncoated and coated SN-282 and AS-800 silicon nitride grades

Specimen condition	Failure statistics, %			
	Surface flaw	Edge/ corner flaw	Volume flaw	Coating/ interface flaw
Uncoated	10	0	90	0
Uncoated/plasma exposed	8	0	92	0
Uncoated/grit blasted	100	0	0	0
Grit blasted/coated	90*	10	0	0
Molten salt etched/coated	90*	10	0	0
Coated	0	52	18	30
Coated/chamfered	0	0	20	80

* Flaws created on the surface of the substrate by surface preparation method before coating.

Table II   Physical, thermal and mechanical properties of silicon nitride substrate and stand alone layers of EBC coatings [18]

Material	Density ρ_{phy} gm/cc	Density ρ_{Arc} gm/cc	Bend Modulus GPa	Bend Strength MPa	Thermal Expansion, α $10^{-6}/^{\circ}C$	Poisson's Ratio, γ
Substrate Silicon nitride(SN-282)	3.4	3.4	318	595	3.8	0.18
Plasma sprayed coatings Mullite	2.6	2.9	45	28	5.8	0.17
Silicon	2.1	2.3	97	40	4.5	0.21
BSAS	2.7	3.2	32	28	5.6	0.19

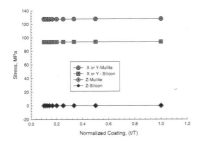 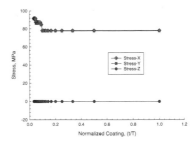

Fig. 6. Predicted thermal residual stresses in x, y, and z directions of the coating as a function of distance (t) from the substrate for (a) 75 μm thick mullite or silicon, and (b) 75 μm silicon/75 μm (mullite+BSAS)/75 μm BSAS. In the figure, t and T represent incremental and total thickness of the coating, respectively.

DISCUSSION

In general, the strength of monolithic ceramics is controlled by the largest flaw normal to the loading direction. Knowing the size of the largest flaw, and the elastic modulus and fracture toughness values of the material, the fracture strength can be calculated, or knowing the fracture strength, the strength critical flaw size can be calculated from the Griffith's equation. Substituting fracture toughness values of 5.5 and 8 MPa.m$^{0.5}$ [19], and elastic modulus values of 310 and 318 GPa [18] in Griffith's equation, the calculated the strength critical flaw size for as-produced SN-282 and AS-800 grades of silicon nitride are ~65 and 90 μm, respectively. In the case of coated silicon nitride, prediction of strength critical flaw is complicated by residual stress in the coating. The coating process itself can create new flaws which are much larger than the strength critical flaw in the as-produced substrate or thermal residual stress generated during deposition of the coating can change stress state on the substrate and activate the largest flaw present on its surface and cause failure. According to the Griffith's equation, to reach 50% loss in strength in plasma spray coated SN-282 and AS-800 grades of silicon nitride, the predicted size of the strength controlling flaw is ~250 and 350 μm, respectively. However, the fractographic analysis of the coated silicon nitride shows much smaller flaw size (~100 to 150 μm) and the origin of these flaws varies depending on the surface preparation of the specimens before and after the coating process. This suggests that various steps used in the coating process cause strength degradation and some of the strength degrading mechanisms may be overlapping, resulting in approximately the same net loss in strength. By proper care, strength degradation caused by surface preparation methods and edge effects can be eliminated. However, even in a well controlled coated specimen, strength degradation is observed and invariably the failure occurs from a flaw at the coating/substrate interface or within the coating. The fact that the strength of coated silicon nitride is relatively independent of the coating thickness or the composition of deposited layers and that most of the failure in well controlled coated substrates occur in the vicinity of the coating/substrate interface also suggests activation of preexisting flaw(s) on the substrate surface or coating defects act as stress raisers due to imposed tensile thermal residual stresses. Indeed, FEM calculation does indicate ~80 to 100 MPa tensile thermal residual stresses in the coating. Such a large thermal residual stress can activate pre existing surface flaws or flaws created during plasma deposition, even though these flaws are well below the size of the strength critical flaws in the uncoated silicon nitride substrate. Therefore, based on current data, it appears that strength degradation in coated silicon nitride is due to residual stress activated preexisting flaws or the flaws created by the plasma deposition on the substrate surface. One way of reducing this effect is to deposit a functionally graded porous silicon carbide or silicon nitride bond coat layer on the substrate by slurry or sol gel coating before plasma deposition. Thus, cracks formed in the EBC are deflected at the bond coat/ intermediate coat interface, relieving residual stress effect.

SUMMARY OF RESULTS
Room temperature flexural strength and strength controlling flaws in two high temperature grades of silicon nitrides that were coated with single layer of PS silicon or mullite and multi layered EBC were determined. The maximum thermal residual stresses generated in the PS coatings were modeled using a finite element program. Major findings are the following.
  (1) Surface preparation methods such as grit blasting or molten salt etching used for better adherence plasma spray cause significant damage to the substrate surface and hence strength degradation.
  (2) Coated specimens need to be re-chamfered to avoid edge and corner failures.
  (3) As-machined silicon nitride bars that were coated and chamfered also show strength degradation.
  (4) Tensile residual stresses in the coating, activation of preexisting flaws and well bonded splats of plasma sprayed coating acting as stress raisers are the reasons for the strength degradation of coated silicon nitride.

CONCLUSION
Silicon nitride substrate coated with plasma sprayed EBC shows strength degradation. By developing a functionally graded bond coat layer deposited by a process other than plasma spray, it may be possible to deflect crack within the coating and reduce the tensile residual stress effect of EBC on silicon nitride substrate and thus avoid strength degradation.

REFERENCES

[1] W.D Brentnall, M. van Roode, P.F. Norton, S. Gates, J.R. Price, O. Jimenez, and N. Miriyala, "Ceramic Gas Turbine Development at Solar Turbines Incorporated," p155-192, V 1, eds M. van Roode, M.K. Ferber, and D.W. Richerson, The American Society of Mechanical Engineers, Three Parks Avenue, New York, NY 10016, 2003.
[2] T. Itoh, Y.Yoshida, S. Sasaki, M. Sasaki, and H.Ogita, "Japanese Automotive Ceramic Gas Turbine Development," p283-303, V 1, eds M. van Roode, M.K. Ferber, and D.W. Richerson, The American Society of Mechanical Engineers, Three Parks Avenue, New York, NY 10016, 2003.
[3] E.A. Gulbransen and S.A. Jansson, "The High-Temperature Oxidation, Reduction, and Volatilization Reactions of Silicon and Silicon Carbide," *Oxidation of Metals*, 4[3] p181-201, (1972)

[4] P.J. Jorgensen, M.E. Wadsworth, and I.B. Cutler,"Oxidation of Silicon Carbide," *J. Am. Ceram. Soc.*, **42** (12): p613-616 (1959).

[5] J.L. Smialek, R.C. Robinson, E.J. Opila, D.S. Fox, and N.S. Jacobson, "SiC and Si_3N_4 Recession Due to SiO_2 Scale Volatility under Combustor Conditions," *Adv. Composite Mater*, **8** [1], 33-45 (1999).

[6] K.N. Lee, *Surface Coating Technology*, p133-134, 1-7, 2000.

[7] K.N. Lee, D.S. Fox, J.I. Eldridge, D. Zhu, R.C. Robinson, N.P. Bansal, and R.A. Miller, *J. Am. Ceram. Soc.*, **86** (8): p1299-1306, 2003.

[8] K.N. Lee, D.S. Fox, and N.P. Bansal,"Rare earth silicate environmental barrier coatings for SiC/SiC composites and Si_3N_4 ceramics," *J. Eur. Ceram. Soc.*, **25** (10): p1705-1715, 2005.

[9] R.T. Bhatt, G.N. Morscher, and K.N. Lee, "Influence of EBC Coating on Tensile Properties of MI SiC/SiC Composites," Proceedings of PACRIM conference, 2005

[10] H.T. Lin, M.K. Ferber, A.A. Werezak, and T.P. Kirkland,"Effects of Material Parameter on Strength of EBC Coated Silicon Nitride" EBC Workshop (2003).

[11] T. Bhatia, V.R.Vedula, H.E.Eaton, E.Y.Sun, J.H.Holowczak, and G.L.Linsey,"Development and Evaluation of Environmental Barrier Coatings for Si-based Ceramics", ASME Paper GT2004-54092.

[12] W.K.Tredway, J. Shi, J.H.Holowczak, V.Vedula, C.Bird, S.Ochs, L.Bertuccoioli, and D. Bombara,"Design of Ceramic Components for an Advanced Micro-Turbine Engine" ASME Paper GT2004-54205.

[13] S. M. Zemskova, H. T. Lin, M. K. Ferber and A. J. Haynes, "Performance of CVD Mullite Coatings on Silicon Nitride under High Temperature High Load Conditions" Key Engineering Materials Vol. 287 (2005) pp. 457-470.

[14] H.T. Lin, M.K. Ferber, T.P. Kirkland and S.M. Zemskova,"Dynamic Fatigue of CVD-Mullite Coated SN88 Silicon Nitride" ASME Paper GT2003-38919.

[15] D.P. Butt, J.J. Mecholsky, M. van Roode, and J.R. Price,"Effects of Plasma Sprayed Ceramic Coatings on the Strength Distribution of Silicon Carbide Materials" J. Am. Ceram. Soc., 73[9] 2690-2696 (1990)

[16] D. Jayaseelan, S.Ueno and T.Ohji,"Sol-gel Synthesis and Coating of Nano-Crystalline Lu2Si2O7 on Si3N4 Substrate" Materials Chemistry and Physics, 84, 192-195 (2004).

[17] K.N. Lee, R.A. Miller, and N.S. Jacobson,"New Generation of Plasma-Sprayed Mullite Coatings on Silicon-Carbide," *J. Am. Ceram. Soc.*, **78** (3) p705-710 (1995).

[18] R.T. Bhatt and D.S. Fox, "Thermo-Mechanical Properties and Durability of Stand Alone Constituent Layers of BSAS and RES based EBCs" in preparation

19 "Ceramic Gas Turbine Design and Test Experience," Volume I of Progress In Ceramic Gas Turbine Development, M. van Roode, M.K. Ferber, D.W. Richerson, eds., The ASTM, Three Park Avenue, New York, NY 10016, 2003

EFFECT OF MICROSTRUCTURE CHANGE OF APS-TBCS ON THE INTERFACIAL MECHANICAL PROPERTY UNDER SHEAR LOADING

Makoto Hasegawa and Hiroshi Fukutomi
Division of Materials Science and Engineering, Yokohama National University
79-5 Tokiwa-dai, Hodogaya-ku
Yokohama, 240-8501, Japan

ABSTRACT
Pushout tests were carried out on APS TBCs which were heat treated in a vacuum and heat exposed in an air in order to understand the effect of microstructural change on interfacial mechanical properties. Heat treatment and heat exposure were conducted at 1413 K from 10 to 50 hours and at 1173 and 1413 K from 10 to 200 hours, respectively. Yield stress of the BC layer estimated by Vickers hardness decreases and the interfacial shear strength increases with increasing heat treatment time. The TGO thickness increases and yield stress of the BC layer decreases with the increase in heat exposure time independently of the heat exposure temperature. Shear strength decreases monotonically with the increase in heat exposure time. As for the specimens heat exposed at 1423 K, interfacial shear strength decreases gradually than the specimen heat exposed at 1173 K.

INTRODUCTION
Thermal barrier coating systems (TBCs) have been widely used to improve the durability of hot section components in gas turbines blades [1-3]. TBCs usually consist of outer oxide ceramics thermal barrier coating (TBC) layer and inner intermetallic bond coat (BC) layer in order to protect the Ni-base supperalloy from high temperature and oxidation. It is reported that delamination of TBC layer initiates at mode I loading condition under a thermally loaded condition. Then, the delamination of TBC layer proceeds by mode II loading condition [2]. Durability of TBCs depends on the interaction between crack driving force and the resistance of crack propagation along the interfaces. Therefore, evaluation of interfacial fracture toughness of TBC layer under mode II loading condition is an important subject for the research.

Recently, indentation, wedge impression, barb and pushout test methods have been developed to evaluate the interfacial fracture toughness under mode II loading condition [4-9]. Evaluation of interfacial fracture toughness of heat exposed and thermally cycled EB-PVD TBCs was performed by barb and pushout test method [6-9]. Microstructure change of TBCs under heat exposure and thermal cycle condition has never considered on the interfacial fracture toughness, though the properties of constitutes and entire system of TBCs depend on the microstructure of TBCs [10-13]. Interfacial mechanical properties of TBCs in different microstructure in mode II loading condition is still not well known. The present study focused on the effect of microstructural change of TBCs on interfacial shear strength.

EXPERIMENTAL PROCEDURE
The TBC systems used in this experiment have been fabricated by plasma spraying process. Bond coat material was a NiCoCrAlY alloy (Co22 Ce17 Al12.5 Y0.6 and balance Ni in mass%) that was coated on a nickel base superalloy substrate (Inconel738, Co8.5 Cr16 Al3.5

Ti3.5 W2.6 Mo1.8 Nb0.9 and balance Ni in mass%) by a low-pressure plasma spraying process. Thermal barrier coating layer of an 8 mass% Y_2O_3 partially stabilized ZrO_2 was coated on BC layer by an air plasma-spraying process. The thickness of TBC, BC layers and substrate was ~250, ~100 μm and 3.2 mm, respectively. The coated material was heat treated in a vacuum at 1413 K from 10 to 50 hours. As-sprayed TBCs were heat exposed in an air at 1173 and 1423 K from 10 to 200 hours. Microstructure change during heat treatment and heat exposure was characterized on the polished transverse section of the TBCs using optical microscope and scanning electron microscope. Yield stress of BC layer in as-sprayed, heat treated and heat exposed TBCs was evaluated by Vickers hardness measurement based on $\sigma_y = HV/3$.

Pushout test method was used to measure interfacial shear strength. Substrate side of the TBCs was bonded by epoxy-based adhesive and polished to form pushout specimen. Figure 1 shows the schematic configuration of the pushout test method. The size of the specimen was 5 mm in hight, c, 4 mm in width, w, and 6.4 mm in thickness, 2h. The pushout test was done in air at room temperature using screw-driven type testing machine with the constant cross head speed of 0.2 mm/min. After pushout tests, delaminated areas were observed by OM and SEM.

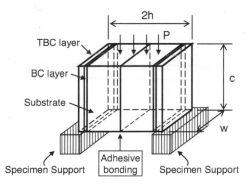

Fig. 1 Schematic configuration of pushout test method.

RESULTS AND DISCUSSION

Microstructural change during heat treatment and heat exposure

Figure 2 shows entire view of the polished transverse section of the as-sprayed APS-TBC system. TBC layer, BC layer and substrate are clearly seen. Pores and inter-splat boundaries are observed in the TBC layer as black spots and lines, respectively. The BC layer seems to be a single phase. However, high-magnification view shows that the BC layer composes of fine mixture of bright γ' phase and dark β phase [14-16]. Microstructure of BC layer, TBC-BC and BC-substrate interfaces after heat treatment at 1413 K for 10 and 50 hours are shown in Fig. 3. With the increase in heat treatment time, the BC separates into 2 different zones. Zone adjacent to the TBC layer composed of bright and dark contrast phases and zone adjacent to the substrate composed of bright contrast phase. The thickness of zone adjacent to the TBC layer decreases and the thickness of zone adjacent to the substrate increases with the increase in heat treatment time. Further, the size of bright and dark contrast phases in zone adjacent to TBC layer increases with increasing heat treatment time. The value of Vickers hardness in BC layer near TBC layer

Fig. 2 SEM micrograph of as-sprayed APS-TBCs observed from transverse section.

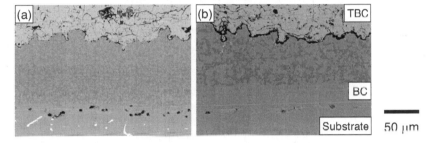

Fig. 3 SEM micrographs of heat treated APS-TBCs seen from transverse section. Heat treated at 1413 K for 10 (a) and 50 (b) hours.

decreased from 369 to 331 with the increase in heat treatment time.

Figure 4 shows the microstructure of BC layer heat exposed at 1173 K for 10 and 200 hours (Fig. 4(a) and (b)) and at 1423 K for 10 and 200 hours (Fig. 4(c) and (d)). TGO layer forms at TBC-BC interface during the heat exposure. The thickness of TGO layer increases with increasing heat exposure temperature and time. After heat exposure, dark β phase disappeared and the zones that existed adjacent to the TBC layer and substrate become a bright γ' single phase. The thickness of zones adjacent to the TBC layer and substrate increases with the increase in heat exposure temperature and time. It is also seen that dark contrast phase disappeared from the BC layer and entire BC layer became a single phase with bright contrast [13]. The Vickers hardness decreases with the increase in heat exposure time independently of heat exposure temperature. The Vickers hardness of BC layer near TBC layer which were heat exposed for 0, 10 and 200 hours at 1173K was 369, 366 and 358, respectively. The hardness of BC layer after exposure at 1423 K for 10 and 200 hours was 324 and 302, respectively.

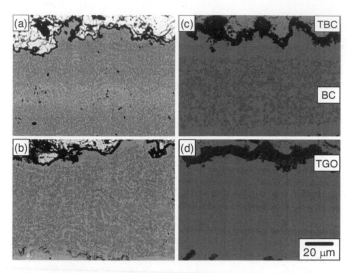

Fig. 4 SEM micrographs of heat exposed APS-TBCs observed from transverse section. Heat exposed at 1173 K for 10 (a) and 200 (b) hours and at 1423 K for 10 (c) and 200 (d) hours.

Effect of BC layer microstructure in interfacial sear strength

Figure 5 shows the typical load-displacement curve obtained by pushout test. Nonlinearity near the origin is observed because of compliance change of entire loading system. After the nonlinear behavior, linear regime appeared. Interfacial shear strength was evaluated as the maximum load appearing in the load-displacement curve. Interfacial shear strength in different heat treatment time is shown in Fig. 6. Interfacial shear strength increases with the

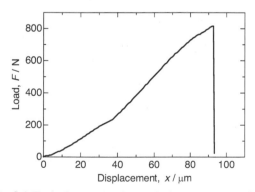

Fig. 5 A Typical example of load-displacement curve of pushout test.

increase in heat treatment time. As for pushout test method, it is assumed that entire stored strain energy in a pushout test specimen at a maximum load is converted to form new surface in TBC-BC interface [9, 10, 16]. Interfacial shear strength is related to the interfacial fracture toughness. Interfacial fracture toughness increases with increasing process zone size ahead the decohesion front and the process zone size increases with decreasing yield stress of the material [17, 18]. Thus, the increase of interfacial shear strength may be due to the decrease in yield stress of BC layer that may increase the process zone size ahead the decohesion front of TBC-BC interface. Figure 7 shows the interfacial shear strength in a different heat exposure temperature and time. Interfacial shear strength decreases monotonically with the increase in heat exposure time. When the specimens were heat exposed at 1423 K, interfacial shear strength decreases gradually than the strength from the specimen heat exposed at 1173 K. The monotonic decrease of

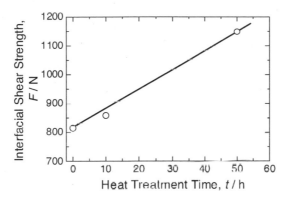

Fig. 6 Change in interfacial shear strength in different heat treatment time.

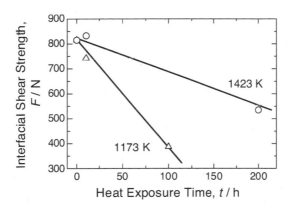

Fig. 7 Change in interfacial shear strength in different heat exposure time.

interfacial shear strength may be due to the increase of residual stress by the formation and growth of TGO layer and the formation of spinels [11, 19]. The change of gradient during decrease may be due to the interaction between increase of residual stress and the decrease of yield stress of the BC layer.

CONCLUSIONS

Microstructural change of BC layer and measurement of interfacial shear properties has been studied by heat treated and heat exposed air plasma-sprayed TBCs. BC layer consists of fine mixture of bright γ' phase and dark β phase. The size of both phases increase with an increase in heat treatment time. Dark β phase adjacent to the TBC layer and substrate disappears during the heat exposure. Yield stress of BC layer estimated by Vickers hardness decreases and interfacial shear strength increases with the increase in heat treatment time. This may be due to the decrease in yield stress of BC layer that may increase the process zone size ahead the decohesion front. Yield stress of BC layer decreases with the increase in heat exposure time independently of heat exposure temperature. Interfacial shear strength decreases monotonically with the increase in heat exposure time. As for the specimens heat exposed at 1423 K, interfacial shear strength decreases gradually than the specimen heat exposed at 1173 K. The change of gradient during decrease may be due to the interaction between increase of residual stress by the formation and growth of TGO layer and the decrease of yield stress of the BC layer.

REFERENCES

[1] S. M. Meier and D. K. Gupta,"The Evolution of Thermal Barrier Coatings in Gas Turbine Engine Applications," *Trans. ASME J. Eng. Gas Turbines Power* **116**, 250-257 (1994).

[2] A. G. Evans, D.R. Mumm, J. W. Hutchinson, G. H. Meier and F. S. Pettit,"Mechanisms Controlling the Durability of Thermal Barrier Coatings," *Prog. Mater. Sci.* **46**, 505-553 (2001).

[3] N. P. Padture, M. Gell and E. H. Jordan,"Thermal Barrier Coatings for Gas-Turbine Engine Applications," *Science*, **296** [5566] 280-284 (2002).

[4] A. Vasinonta and J. L. Beuth,"Measurement of Interfacial Toughness in Thermal Barrier Coating Systems by Indentation,"*Eng. Fracture Mech.* **68**, 843-860 (2001).

[5] M.R.Begley, D. R. Mumm, A. G. Evans and J. W. Hutchinson,"Analysis of A Wedge Inpression Test for Measuring The Interface Toughness Between Films/Coatings and Ductile Substrate," *Acta Mater.*, **48** 3211-3220 (2000).

[6] S. Guo, D. R. Mumm, A. M. Karlsson and Y. Kagawa,"Measurement of Interfacial Shear Mechanical Properties in Thermal Barrier Coating Systems by a Barb Pullout Method," *Scripta Mater.*, **53** 1043-1048 (2005).

[7] S. Guo, Y. Tanaka and Y. Kagawa,"Effect of Interface Roughness and Coating Thickness on Interfacial Shear Mechanical Properties of EB-PVD Yttria-Partially Stabilized Zirconia Thermal Barrier Coating Systems," *J. Euro. Ceram. Soc.*, **27** 3425-3431 (2007).

[8] S. S. Kim, Y. F. Liu and Y. Kagawa, "Evaluation of Interfacial Mechanical Properties under Shear Loading in EB-PVD TBCs by The Pushout Method," *Acta Mater.*, **55** 3771-3781 (2007).

[9] M. Tanaka, Y. F. Liu, S. S. Kim and Y. Kagawa,"Delamination Toughness of Electron Beam Physical Vapor Deposition (EB-PVD) Y_2O_3-ZrO_2 Thermal Barrier Coatings by The Pushout Method: Effect of Thermal Cycling Temperature," *J. Mater. Res.*, **23** 2382-2392 (2008).

[10]S. Guo and Y. Kagawa,"Young's Moduli of Zirconia Top-Coat and Thermally Grown Oxide in a Plasma-Sprayed Thermal Barrier Coating Sysrwm," *Scripta Mater.*, **50** 1401-1406 (2004).

[11]A. Shinmi, M. Hasegawa, Y. Kagawa, M. Kawamura and T. Suemitsu,"Change in Microstructure of Plasma Sprayed Thermal Barrier Coating by High Temperature Isothermal Heat Exposure," *J. Japan Inst. Metals,* **69** 67-72 (2005).

[12]M. Tanaka, M. Hasegawa, A. F. Dericioglu and Y. Kagawa,"Measurement of Residual Stress in Air Plasma-Sprayed Y_2O_3-ZrO_2 Thermal Barrier Coating System using Micro-Raman Spectroscopy," *Mater. Sci. Eng.*, **A419** 262-268 (2006).

[13]M. Hasegawa and Y. Kagawa,"Microstructural and Mechanical Properties Change of a NiCoCrAlY Bond Coat with Heat Exposure Time in Air Plasma-Sprayed Y_2O_3-ZrO_2 TBC Systems," *Int. J. Appl. Ceram. Technol.*, **3** 293-301 (2006).

[14]C.Mennicke, D. R. Mumm and D. R. Clarke,"Transient Phase Evolution During Oxidation of a Two-Phase NiCoCrAlY Bond Coat," *Z. Metallk.*, **90** 1079-1084 (1999).

[15]T. Rehfeldt, G. Schmacher, R. Vaβen and R. P. Wahi,"Order-disorder Transformation in a NiCoCrAlY Bond Coat Alloy at High Temperature," *Scripta. Mater.*, **43** 963-968 (2001)

[16]M. Hasegawa, T. Endo and H. Fukutomi,"The Effect of Microstructure Change of Bond Coat Layer in Air Plasma-Sprayed Thermal Barrier Coating System on Interfacial Mechanical Property under Shear Loading," *J. Japan Inst. Metals*, **73** 802-808 (2009).

[17]Y. Wei and J. W. Hutchinson,"Nonlinear Delamination Mechanics for Thin Films," *J. Mech. Phys. Solids*, **45** 1137-1159 (1997).

[18]V. Tvergaard and J. W. Hutchinson,"The Influence of Plasticity on Mixed Mode Interface Toughness," *J. Mech. Phys. Solids*, **41** 1119-1135 (1993).

[19]M. Tanaka, M. Hasegawa and Y. Kagawa,"Detection of Micro-Damage Evolution of Air Plasma-Sprayed Y_2O_3-ZrO_2 Thermal Barrier Coating through TGO Stress Measurement," *Mater. Trans*, **47** 2515-2517 (2006).

EFFECT OF THERMAL GRADIENT ON THE THROUGH-THICKNESS THERMAL CONDUCTIVITY OF PLASMA-SPRAYED TBCS

Yang Tan[*,†], Jon Longtin and Sanjay Sampath[†]
Center for Thermal Spray Research, Stony Brook University
Stony Brook, New York, USA

Dongming Zhu[†]
NASA Glenn Research Center
Cleveland, Ohio, USA

ABSTRACT
    The harsh thermal environment in gas turbines, including elevated temperatures and high heat fluxes, induces significant thermal gradients in ceramic thermal barrier coatings (TBCs), which are used to protect metallic components. However, the thermal conductivity of plasma-sprayed TBC increases with exposure at high temperatures mainly due to sintering phenomena and possible phase transformation, resulting in potential thermal runaway issues. An analytical thermal model, as well as coating thermal conductivity data which are experimentally obtained, are used to determine the coating through-thickness temperature profile and effective thermal conductivity under gradient conditions at high temperatures. High heat flux rig tests are then performed on TBCs to evaluate coating thermal behavior under temperature gradient closed to service conditions. This combined approach provides a new sintering model and allows for assessment of temperature gradient effects on the thermal performance of plasma-sprayed TBCs.

I. INTRODUCTION

    Thermal barrier coatings (TBCs) based on yttria stabilized zirconia (YSZ)[1] protect metallic turbine components from elevated temperatures. A key attribute of plasma-sprayed TBCs is their microstructure, which is comprised of a myriad array of pores and cracks rendered during the coating deposition process, which further reduces the already low intrinsic thermal conductivity of bulk YSZ.[2,3] In recent years, economic and environmental pressures have driven turbine operational temperatures higher to increase thermodynamic efficiency. The sintering issue of the as-deposited TBCs has thus become significant,[4] since sustained exposures to high temperatures result in the dynamic evolution of the porous ceramic microstructure.

    Figure 1 shows a schematic of a TBC system comprised of a plasma-sprayed ceramic top-coat, a thin thermally-sprayed metallic bond coat, and a superalloy component. Operations at elevated temperatures cause sintering, which leads to increased thermal conductivity[5-7] as well as stiffening of the coating[8,9] with concomitant micro-delamination and spallation.[10] Furthermore, plasma-sprayed YSZ is in the metastable tetragonal phase in the as-deposited state, but can transform to stable cubic and monoclinic phases at temperatures greater than 1200 °C for long-term exposures.[11-13] In the temperature range of 1000–1200 °C, sintering occurs with microstructure evolution that causes undesirable increases in coating stiffness and thermal conductivity, even without major phase change.[13,14] To illustrate, the cross-sectional microstructure images of a plasma-sprayed coating are shown in Figure 2 for as-deposited and heat treated at 1100 °C and 1200 °C for 225 hours. As a result of heat treatment, the fine interfaces disappear, which reduces the defects present in the coating and consequently increases the coating thermal conductivity.

    During operation, the turbine is under thermal gradient conditions, such as in combustors, vanes and blades, especially those incorporate cooling on the metallic components. Because of the intrinsically low thermal conductivity, the TBC on those components will experience the largest temperature gradient, which is oriented along the coating through-thickness direction, i.e., normal to the coating surface. The depth-dependent temperature distribution can potentially results in localized microstructure and thermal conductivity changes within the coating. Therefore, the temperature distribution within the coating will subsequently change, resulting in a time-varying coating thermal profile. Experimental assessment of the temperature distribution is challenging and modeling approaches are needed to describe the gradient effects.

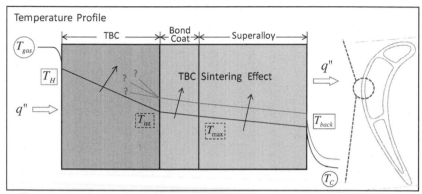

Figure 1. A schematic of a plasma-sprayed TBC system and the temperature distribution profiles for both before and after service.

This study combines the experimental results from measurements on various monolithic TBC specimens subjected to discrete temperature exposure, to enable a new method to predict coating thermal conductivity for arbitrary thermal exposure history. Based on the thermal analysis, the through-thickness temperature distribution and coating effective thermal conductivity are assessed under service conditions. Field tests, which were under high heat flux and enhanced with in-situ measurement of coating effective thermal conductivity, were performed at various conditions for multiple TBCs. The results combined with model-based analysis provide an overall sintering map on thermal behavior of plasma-sprayed YSZ thermal barrier coatings.

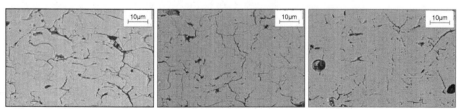

Figure 2. SEM cross-sectional microstructure images for plasma-sprayed TBC, as-deposited (left); after heat treatment at 1100 °C for 225 hours (middle); after at 1200 °C for 225 hours (right).

II. THEORY: ONE-DIMENSIONAL HEAT CONDUCTION

Heat transfer in plasma-sprayed TBC system is complex at the high service temperatures seen by modern TBCs. Conduction dominates, and is also temperature dependent. Radiation is non-negligible.[3] At high temperatures, bulk YSZ becomes semi-transparent to radiative photon heat transfer.[15-17] On the other hand, plasma-sprayed coatings are two-phase composites with ceramic and ambient gases within the pores. A typical coating has a large number of interlaminar cracks and micro- or nano-sized interfaces, which provide significant phonon and photon scattering and reduce the overall TBC thermal conductivity. The convective heat transfer inside the coating is insignificant due to the small size of the defects along the heat flux direction, which is usually below several microns for plasma-sprayed coatings.[18]

The general steady-state transient heat conduction equation without heat generation in one-dimension (1-D) can be expressed as:

$$\frac{d}{dx}\left(k\frac{dT}{dx}\right) = 0 \tag{1}$$

where T is the temperature, x is the distance into the coating and k is the thermal conductivity of the material. For TBC systems, since the overwhelming majority of heat flux is along the coating through-thickness direction, a one-dimensional (1-D) analysis is well justified. Note that thermal conductivity is *not* a constant, but rather depends on both current coating temperature and the prior temperature history of the coating. The heat flux, q'' can be related to the temperature gradient dT/dx:

$$q'' = -k(dT/dx) \tag{2}$$

A schematic of a 1-D model is drawn in Figure 3, which consists of a coating of thickness H, split into n layers with equal thickness $d = H/n$. Here x denotes the distance into the coating, with $x = 0$ defined at the top of the coating. The thermal conductivity of layer i ($i = 1, 2, ..., n$) is denoted by k_i. The temperatures for the top-surface is denoted by T_0, and the temperature for the interface below each layer are denoted by T_i. By using k_{eff} to denote the effective thermal conductivity, Eq.2 applies to the whole system,

$$q'' = -k_{eff}\,(T_n - T_0)/H \tag{3}$$

as well as to each layer:

$$q'' = k_i(T_{i-1} - T_i)/d, \quad (i = 1, 2, ..., n) \tag{4}$$

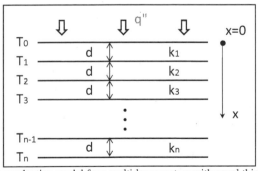

Figure 3. A 1-D heat conduction model for a multi-layer system with equal thickness for each layer.

The interface temperatures T_i can be solved as:

$$T_i = T_0 - q''d \sum_{m=1}^{i} \left(\frac{1}{k_m}\right), \quad (i = 1, 2, \dots, n) \tag{5}$$

The temperature in any arbitrary location can be derived by defining dimensionless location and temperature, $x^* = x/H$ and $T^* = (T - T_n)/(T_0 - T_n)$, respectively:

$$T^* = \left[\frac{i - nx^*}{k_i} + \sum_{m=i+1}^{n} \left(\frac{1}{k_m}\right)\right] / \left[\sum_{m=1}^{n} \left(\frac{1}{k_m}\right)\right], \quad \left(\frac{i-1}{n} < x^* < \frac{i}{n}, \; i = 1, 2, \dots, n\right) \tag{6}$$

The ranges for the dimensionless location and temperature are $0 \leq x^* \leq 1$ and $0 \leq T^* \leq 1$, respectively. Eq.6 enables calculation of temperature distribution profile for an entire multi-layer system with known heat flux and thermal conductivity for each layer. The effective thermal conductivity for the whole system is a function of each layer's thermal conductivity (assuming each layer has uniform thickness):

$$k_{eff} = \frac{n}{\sum_{m=1}^{n} \left(\frac{1}{k_m}\right)} \tag{7}$$

The above equations (3–7) represent an analytical solution to multi-layer systems given discrete values of thermal conductivity of each layer.

In reality, a system may have continuously varying thermal conductivity. In such as case, analytical solutions can be obtained by integration. Results are provided here for two simplified cases with known thermal conductivity relationships.

Case (i): The thermal conductivity has a linear relationship with dimensionless location variable x^*, $k = C_o + C_1 x^*$. Temperature distribution in terms of location:

$$T^* = \frac{\ln\left(\frac{C_o + C_1 x^*}{C_o + C_1}\right)}{\ln\left(\frac{C_o}{C_o + C_1}\right)} \tag{8}$$

And the effective thermal conductivity becomes:

$$k_{eff} = C_1 / \ln\left(\frac{C_o}{C_o + C_1}\right) \tag{9}$$

Case (ii): Thermal conductivity has a linear relationship with dimensionless temperature variable T^*, $k = C_o' + C_1' T^*$. The temperature distribution is in an implicit form:

$$x^* = \frac{-1}{1 + 2C_o'/C_1'}(T^* - 1)(T^* + 1 + 2C_o'/C_1') \tag{10}$$

with effective thermal conductivity

$$k_{eff} = C_0' + \frac{1}{2}C_1' \tag{11}$$

Therefore, for 1-D conductive heat transfer problems, once the thermal conductivities are known, or if they can be reliably estimated or predicted, a solution for the temperature and heat flux can be obtained with the above approach. For some complex systems, e.g., those including radiative heat transfer, numerical simulations may be needed, e.g., using finite element analysis (FEA). For thermal conductivity data of plasma-sprayed TBCs, however, experimental and empirical methods should be satisfactory to obtain, estimate and predict the coating thermal behavior. This work will concentrate on the continuous model because it is more relevant to YSZ based TBCs.

III. EFFECT OF ISOTHERMAL HEAT TREATMENT

In turbine engines, operating conditions, such as temperature fluctuation, cycling environment, etc., can vary significantly, which make it extremely difficult to evaluate and quantify the thermal behavior for arbitrary parameters. Instead, investigations on the sintering effects by either isothermal heat treatment[19] or field-tested coating systems under high heat flux have been conducted.[20] Thermal cycling tests with rapid heating/cooling cycles also become attractive since they are relevant to coating performance. The coating can experience cracking and delamination under thermal cycles,[4; 21] which can cause a desirable decrease in thermal conductivity, but may lead to mechanical failures as well.

Isothermal heat treatment, in particular, has several advantages, including convenience, high temperature accuracy, ease of temperature control and cost-effective operation. Studying TBCs under isothermal exposure provides a first approximation to the thermal behavior of coating systems, and enables modeling and prediction of coating thermal conductivity for various thermal loads.

3.1. Thermal Conductivity of Plasma-Sprayed TBCs

In the case of plasma-sprayed TBCs, the thermal conductivity k is not only temperature-dependent, but also depends on the coating thermal history as it undergoes sintering.[5] The thermal conductivity then is a function of these three parameters.

$$k^* = k/k_{as} = k^*(T, t, \theta) \tag{12}$$

where k_{as} is the coating thermal conductivity as-deposited, T is the current temperature of the coating, θ is the sintering temperature and t is the exposure time. To eliminate the variation of thermal conductivity of samples from different locations in the same coating, which can be up to 3-6%,[22] a normalized thermal conductivity k^* is used. If the coating temperature is above 1000 °C where sintering occurs, and the temperature fluctuation is small, θ can be expressed as follows, during a certain time range, e.g., from t_1 to t_2:

$$\theta = \frac{1}{t_2 - t_1} \int_{t_1}^{t_2} T dt \tag{13}$$

To describe the thermal conductivity in terms of actual temperature and the heat treatment history, an empirical method is introduced in the following section.

3.2. Thermal Conductivity Predictions

The Larson-Miller parameter (LMP) was initially used to describe the behavior of rupture and creep stresses.[23] Since the thermal conductivity increase of TBCs in terms of exposure time and temperature similarly exhibits a 'creep-like' behavior, researchers have used this parameter to equate time and temperature, with change in thermal conductivity for TBCs.[6; 19; 20] It can be expressed as:

$$LMP = \theta(\ln t + 20) \tag{14}$$

The advantage of using *LMP* is to combine the two sintering parameters, time *t* and temperature θ, into a single descriptor. The relationships between *LMP* and coating properties can be experimentally or empirically obtained.[19; 20; 24]

Previous studies considered the thermal conductivity–*LMP* relationship at either ambient or a constant temperature. At elevated temperatures, since coating thermal conductivity is temperature-dependent, coating thermal conductivity will depend not only on sintering history, but also on the current coating temperature, as described in Eq. 12. A new parameter is introduced here:

$$LMP_{HT} = C_0 T + LMP \tag{15}$$

C_0 is a constant and can be evaluated for a particular coating by assessing the impact of temperature on thermal conductivity and *LMP*, separately.

Examples are given here for a plasma-sprayed YSZ coating with an as-deposited thermal conductivity of 1.03 ± 0.02 W·m^{-1}·K^{-1}. Samples from the same coating were heat treated in various isothermal conditions and the temperature-dependent thermal conductivities were measured for both as-deposited and sintered samples by using the laser flash technique. The resulting normalized temperature-dependent thermal conductivities are plotted in Figure 4. From the experimental data, the constant in Eq. (15) can be determined as $C_0 = 2.6$. The natural logarithm of the normalized thermal conductivity, $\ln(k^*)$, is plotted in Figure 5 in terms of LMP_{HT}. The figure contains a large quantity of experimental data and a clear trend can be seen that the $\ln(k^*)$ has a linear relationship with LMP_{HT}, which can be expressed as:

$$\ln(k^*) = b_0 + b_1 LMP_{HT} \tag{16}$$

where b_0 and b_1 are constant parameters obtained from a linear regression of the data. In this work, $b_0 = -1.8$, and $b_1 = 4.3 \times 10^{-5}$.

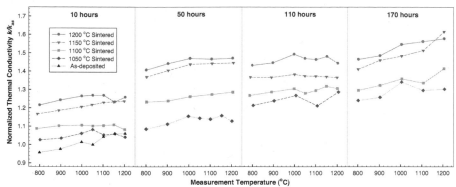

Figure 4. Normalized temperature-dependent thermal conductivity for as-deposited and heat treated TBCs.

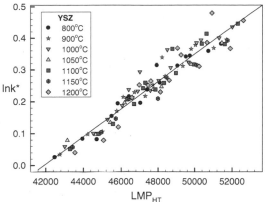

Figure 5. Relationship between normalized thermal conductivity and LMP_{HT} for plasma-sprayed TBCs after various heat treatment.

In a previous work,[24] we obtained LMP constants for each temperature (100, 200, ... , 1200 °C). Equation (16) is its extension that generalizes the results. The parameter C_0 introduces the effect of coating current temperature, which is related to radiation heat transfer. This empirical approach allows one to predict coating thermal conductivity based on a small number of key measurements. Moreover, once the temperature-dependent thermal behavior for a particular coating system has been characterized, i.e., the relationship between current coating temperature and its thermal conductivity is obtained, only additional room-temperature measurements are required to characterize the aging/sintering mechanisms on thermal behavior.

IV. EFFECT OF TEMPERATURE GRADIENT
4.1. Predictions under Temperature Gradient
This section describes how to use the multi-layer or continuous analytical model to simulate a TBC system under temperature gradient conditions, based on measured isothermal heat treatment data for coating samples. When first applying heat flux to the TBC system, the as-deposited coating will reach a quasi-static state, with different locations of the TBC having different temperatures. The TBC will experience different sintering rates throughout the coating due to the non-uniform temperature, and the local thermal conductivity will correspondingly increase at different rates. The system temperature distribution will in turn change in response to the thermal conductivity change. As a result, the local sintering rate will also increase due to the temperature increase. The temperature–thermal conductivity–sintering rates are thus all coupled.

In Eq. 16, for any given LMP_{HT} value, $\ln(k^{*})$ is linear with temperature T. This means at any point in the coating's heat treatment history, the entire temperature distribution profile can be obtained. Consider a single location in the coating at two time points t_1 and $t_1 + \Delta t$, for which the thermal conductivities are k_1^{*} and k_2^{*}, respectively. During time Δt, the coating thermal conductivity will experience sintering. If the sintering time Δt is relatively short, then $\theta \approx T$, and Eq. 16 leads to:

$$\Delta(\ln k^{*}) = \ln k_2^{*} - \ln k_1^{*} = b_1 T \ln\left(\frac{t_1 + \Delta t}{t_1}\right) \qquad (17)$$

Eqs. 8–11 and 16–17 lead to the temperature solution for a coating under temperature gradient conditions, on both the temperature distribution and the effective thermal conductivity over the time. The procedure is as follows:

(i) For an as-deposited coating ($t = 0$), according to the temperature-dependent thermal conductivity shown in Figure 4, the temperature distribution can be calculated by using Eq. 10 (or equivalent equations if k^*-T is non-linear).

(ii) Using Eq. 16, for a small time increment, the location-dependent thermal conductivities, which increase during this period, can be calculated. The new temperature distribution can be calculated based on Eq. 8 (or equivalent equation if k^*-x^* is non-linear). The effective thermal conductivity can be obtained by using Eq. 9 or 11. For non-linear relationships, equivalent equations are needed.

(iii) The time step is incremented and the process repeated until the target time is achieved.

Example with the following boundary conditions was considered: initial TBC top-coat front-surface temperature (T_0 according to Figure 3) 1246 °C and TBC-metal interface temperature (T_n in Figure 3) 920 °C. The as-deposited and heat treated coating thermal conductivities were as taken from Figures 4 and 5. The total heat flux (q") was kept constant. In real application the heat is transferred from hot gases to the components through convection. As sintering of the TBCs occur, even at same operational conditions, q", T_0 and T_n will all slightly change, in the following calculations, q" and T_0 were fixed for simplification and ease of comparison to experiments.

Results were plotted in Figure 6 for the first 50 hours operation, for both effective thermal conductivity of TBC top-coat (k_{eff}) and the difference between T_0 and T_n (denoted by ΔT). The k_{eff} exhibited a "creep-like" behavior, and resulted in significant decreases on the temperature difference ΔT. Consequently the underlying metallic components are similarly exposed to proportionally higher temperatures. In temperature gradient conditions for TBCs bonded with substrates, the substrate material will provide constraints to the top-coat and alter the sintering rate.[25]

The *isothermal* thermal exposure condition at 1246 °C (here $T_0 = T_n$) was also shown in Figure 6 for comparison with the results with temperature gradient. In this case, the effective thermal conductivity increases much faster than that of the gradient condition. It can be concluded that, at operation temperature of TBCs, the presence of the thermal gradient reduces the sinter rate. Therefore, accurate assessment of sintering on TBC thermal conductivity requires consideration of temperature gradient.

4.2. High Heat Flux Test

High temperature and high heat flux tests were performed on a TBC sample at NASA Glenn Research Center using a 3.5 kW CO_2 high heat flux laser. Details on this facility are described elsewhere. The plasma-sprayed TBC sample was deposited on a superalloy substrate (Rene N5) which has been pre-coated with a thin layer of MCrAlY bond coat by low pressure plasma-spraying (LPPS) to improve adhesion.

The total time duration of the high heat flux rig test was roughly 50 hours, with an initial T_0 of 1246°C, and an initial T_n of 920°C, for the as-processed TBC sample. The heat flux q" passed through the TBC was 93.3 ± 0.3 W/cm^2. The high heat flux rig test allowed extraction of thermal conductivity, based on temperature and heat flux data using Eq. (3). Both k_{eff} and ΔT were plotted in Figure 6, along with the predicted value at similar conditions. The coating's effective thermal conductivity increased while the temperature difference decreased, after heat treatment in the presence of thermal gradients, as expected. It can be seen that both experimental and predicted thermal conductivity follow a similar trend, and the rig test results were in good agreement with the model predictions. Additional high heat flux tests were also performed to study thermal barrier coating thermal conductivity in thermal gradients in higher temperature range to validate the model.

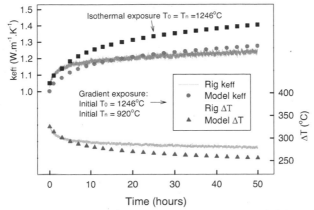

Figure 6. Model and experiments, for 50 hours operation, with initial T_0=1246°C, T_n=920°C. Results are for both k_{eff} and ΔT, with isothermal exposure at maximum temperature shown for comparison.

4.3. Extrapolation to Higher Temperatures

The thermal model were extrapolated to higher temperatures and compared to the high heat flux test, which was performed on an as-processed sample same as in section (4.2), with initial T_0 = 1560°C and T_n = 1120°C. The duration was 50 hours. At higher temperatures, sintering becomes more significant due to microstructure changes. Phase change also become stronger that increases coating thermal conductivity, especially for long-term thermal exposure. Before and after heat treatment, XRD (X-ray diffraction) tests were performed and it turned out there were no significant phase changes found on the coating top-surface. The results for k_{eff} and ΔT were plotted in Figure 7. Comparing model with experiments, both conductivity and temperatures parameters were following the same trend. The model results were in good agreement with the rig test. Heat transfer behavior of TBCs at high temperature is complex, especially considering the radiation heat transfer, which increases non-linearly at high temperatures, as well as phase changes, which can alter the sintering mechanisms. This combined model–experiment approach provides assessment of coating thermal behavior at service conditions under temperature gradient.

To illustrate this method in a broad range, multiple additional tests were performed with various boundary conditions. The above results and previous studies indicated that,[20; 26] during thermal exposure, there is a faster increase in thermal conductivity at early times that slows at later times. We choose 20-hour as a representative time to calculate the relative thermal conductivity increase before and after thermal exposure. Figure 8 plotted the comparison of model and high heat flux tests, on entire coating effective thermal conductivity increase after 20 hours thermal exposure in temperature gradient condition.

Overall, the model seems in good agreement with rig tests. Although it tends to under-predict the conductivity for higher temperatures. This is possibly because of the increased radiation effect[3] and potential phase change issues (monoclinic has quite higher thermal conductivity.[2] Some lower temperature data scattering may be from micro-delamination, which occurs locally but reduces the coating effective thermal conductivity.[8,20,26]

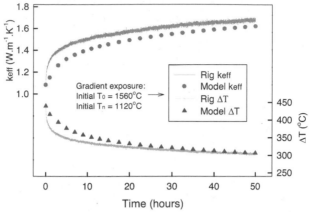

Figure 7. Model and experiments, for 50 hours operation, with initial T_0 of 1560°C and T_n of 1120°C. Results are for both k_{eff} and ΔT.

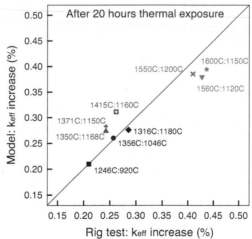

Figure 8. Comparison of model and experiments for various conditions. Results are for k_{eff} increase after 20 hours exposure. Numbers marked are for initial T_0:T_n.

V. CONCLUSIONS

In turbine applications, the TBC systems are subjected to large temperature gradients and in such cases the coating ages and responds differently than that under isothermal conditions at maximum operating temperature. Isothermal aging tests on the entire system typically can only be performed at lower temperatures limited by the metallic substrate temperature capability; while tests under thermal gradient are usually tested at higher coating surface temperature, which would be able to obtain realistically in an engine operation.

This issue can be assessed based on a 1-D thermal analysis and measured data from representative coating samples. The prediction is compared with laser rig tests that produce high heat flux, and provides a previously unavailable estimate of coating thermal behavior exposed to temperature gradients for plasma-sprayed thermal barrier coatings. By comparing with predicted values based on the thermal analysis, the embedded thermocouple provides effective thermal monitoring of the TBC system under high heat flux conditions, which is close to the harsh thermal environments in turbine operations.

The paper provides a new sinter–thermal conductivity model which is validated by experiments, as well as a sintering database for YSZ up to very high temperatures under simulated engine thermal gradients. This work is aiming to improve the understanding of TBC thermal performance under realistic conditions.

ACKNOWLEDGEMENTS

This work was supported by the U.S. National Science Foundation under Award DMI-0428708. The support of the industrial consortium for thermal spray technology is gratefully acknowledged.

FOOTNOTES

* Corresponding author. Email: yangtan@gmail.com (Yang Tan).
† Member, American Ceramic Society

REFERENCES

[1] R. A. Miller, Current Status of Thermal Barrier Coatings - An Overview, *Surface & Coatings Technology*, **30**[1] 1-11 (1987).

[2] P. G. Klemens and M. Gell, Thermal conductivity of thermal barrier coatings, *Materials Science and Engineering a-Structural Materials Properties Microstructure and Processing*, **245**[2] 143-49 (1998).

[3] L. Pawlowski and P. Fauchais, Thermal Transport-Properties of Thermally Sprayed Coatings, *International Materials Reviews*, **37**[6] 271-89 (1992).

[4] J. W. Hutchinson and A. G. Evans, On the delamination of thermal barrier coatings in a thermal gradient, *Surface & Coatings Technology*, **149**[2-3] 179-84 (2002).

[5] R. Dutton, R. Wheeler, K. S. Ravichandran, and K. An, Effect of heat treatment on the thermal conductivity of plasma-sprayed thermal barrier coatings, *Journal of Thermal Spray Technology*, **9**[2] 204-09 (2000).

[6] H. E. Eaton, J. R. Linsey, and R. B. Dinwiddie, The Effect of Thermal Aging on the Thermal Conductivity of Plasma Sprayed Fully Stabilized Zirconia, *Thermal Conductivity 22*, ed. T Tong, Technomic Pub. 289-300 (1994).

[7] J. A. Thompson and T. W. Clyne, The effect of heat treatment on the stiffness of zirconia top coats in plasma-sprayed TBCs, *Acta Materialia*, **49**[9] 1565-75 (2001).

[8] S. R. Choi, D. M. Zhu, and R. A. Miller, Effect of sintering on mechanical properties of plasma-sprayed zirconia-based thermal barrier coatings, *Journal of the American Ceramic Society*, **88**[10] 2859-67 (2005).

[9] S. A. Tsipas, I. O. Golosnoy, R. Damani, and T. W. Clyne, The effect of a high thermal gradient on sintering and stiffening in the top coat of a thermal barrier coating system, *Journal of Thermal Spray Technology*, **13**[3] 370-76 (2004).

[10] A. G. Evans, D. R. Mumm, J. W. Hutchinson, G. H. Meier, and F. S. Pettit, Mechanisms controlling the durability of thermal barrier coatings, *Progress in Materials Science*, **46**[5] 505-53 (2001).

[11] D. R. Clarke and C. G. Levi, Materials design for the next generation thermal barrier coatings, *Annual Review of Materials Research*, **33** 383-417 (2003).

[12] C. G. Levi, Emerging materials and processes for thermal barrier systems, *Current Opinion in Solid State & Materials Science*, **8**[1] 77-91 (2004).

[13]R. W. Trice, Y. J. Su, J. R. Mawdsley, K. T. Faber, A. R. De Arellano-Lopez, H. Wang, and W. D. Porter, Effect of heat treatment on phase stability, microstructure, and thermal conductivity of plasma-sprayed YSZ, *Journal of Materials Science,* **37**[11] 2359-65 (2002).

[14]J. Ilavsky and J. K. Stalick, Phase composition and its changes during annealing of plasma-sprayed YSZ, *Surface & Coatings Technology,* **127**[2-3] 120-29 (2000).

[15]N. M. Balzaretti and J. A. H. Dajornada, PRESSURE-DEPENDENCE OF THE REFRACTIVE-INDEX OF MONOCLINIC AND YTTRIA-STABILIZED CUBIC ZIRCONIA, *Physical Review B,* **52**[13] 9266-69 (1995).

[16]D. L. Wood and K. Nassau, REFRACTIVE-INDEX OF CUBIC ZIRCONIA STABILIZED WITH YTTRIA, *Applied Optics,* **21**[16] 2978-81 (1982).

[17]D. L. Wood, K. Nassau, andT. Y. Kometani, REFRACTIVE-INDEX OF Y2O3 STABILIZED CUBIC ZIRCONIA - VARIATION WITH COMPOSITION AND WAVELENGTH, *Applied Optics,* **29**[16] 2485-88 (1990).

[18]I. O. Golosnoy, A. Cipitria, andT. W. Clyne, Heat Transfer through Plasma-Sprayed Thermal Barrier Coatings in Gas Turbines: A Review of Recent Work, *Journal of Thermal Spray Technology,* **18**[5-6] 809-21 (2009).

[19]F. Cernuschi, L. Lorenzoni, S. Ahmaniemi, P. Vuoristo, and T. Mantyla, Studies of the sintering kinetics of thick thermal barrier coatings by thermal diffusivity measurements, *Journal of the European Ceramic Society,* **25**[4] 393-400 (2005).

[20]D. M. Zhu and R. A. Miller, Thermal conductivity and elastic modulus evolution of thermal barrier coatings under high heat flux conditions, *Journal of Thermal Spray Technology,* **9**[2] 175-80 (2000).

[21]A. G. Evans and J. W. Hutchinson, The mechanics of coating delamination in thermal gradients, *Surface & Coatings Technology,* **201**[18] 7905-16 (2007).

[22]W. Chi, S. Sampath, and H. Wang, Ambient and high-temperature thermal conductivity of thermal sprayed coatings, *Journal of Thermal Spray Technology,* **15**[4] 773-78 (2006).

[23]F. R. Larson and J. Miller, A Time-Temperature Relationship for Rupture and Creep Stresses, *Transactions of the ASME,* **Jul** 765-75 (1952).

[24]Y. Tan, J. P. Longtin, S. Sampath, and H. Wang, Effect of the Starting Microstructure on the Thermal Properties of As-Sprayed and Thermally Exposed Plasma-Sprayed YSZ Coatings, *Journal of the American Ceramic Society,* **92**[3] 710-16 (2009).

[25]A. Cipitria, I. O. Golosnoy, and T. W. Clyne, A sintering model for plasma-sprayed zirconia thermal barrier coatings. Part II: Coatings bonded to a rigid substrate, *Acta Materialia,* **57**[4] 993-1003 (2009).

[26]D. M. Zhu, R. A. Miller, B. A. Nagaraj, and R. W. Bruce, Thermal conductivity of EB-PVD thermal barrier coatings evaluated by a steady-state laser heat flow technique, *Surface & Coatings Technology,* **138**[1] 1-8 (2001).

THERMAL PROPERTY MEASUREMENT FOR THERMAL BARRIER COATINGS
BY THERMAL IMAGING METHOD

J. G. Sun
Argonne National Laboratory
Argonne, IL 60439

Thermal barrier coatings (TBCs) are being extensively used for improving the performance and extending the life of combustor and gas turbine components. TBC thermal properties, thermal conductivity and heat capacity (the product of density and specific heat), are important parameters in those applications. These TBC properties are usually measured by destructive methods, involving separating the ceramic coating layer from the substrate and performing density, specific heat, and thermal diffusivity measurements. Nondestructive evaluation (NDE) methods, on the other hand, allow for direct TBC property measurement on natural TBC samples. Therefore, they can be used for inspecting the quality of as-processed components and monitoring TBC degradation during service. This paper presents a multilayer thermal-modeling NDE method, which analyzes data obtained from pulsed thermal imaging, to determine thermal conductivity and heat capacity distributions over the entire surface of a TBC specimen. The measurement accuracy was investigated and compared with standard destructive measurement data. Experimental results are presented and discussed.

INTRODUCTION

Thermal barrier coatings (TBCs) have been extensively used on hot gas-path components in gas turbines. In this application, a thermally insulating ceramic topcoat (the TBC) is bonded to a thin oxidation-resistant metal coating (the bond coat) on a metal substrate. TBC coated components can therefore be operated at higher temperatures, with improved performance and extended lifetime [1,2]. Because TBCs play critical role in protecting the substrate components, their failure (spallation) may lead to unplanned outage or safety threatening conditions. Therefore, it is necessary to determine the initial condition of as-processed TBCs as well as to monitor the condition change during service to assure their quality and reliability. Because the primary function of a TBC is for thermal insulation, the most important TBC parameters are thermal properties, particularly the thermal conductivity. Experimental methods to accurately and conveniently measure TBC thermal conductivity are still being actively pursued today.

TBC conductivity can be measured by several methods. The most reliable and commonly used method is laser flash method on stand-alone TBC coated specimens [3], which is a special case of the two-sided thermal-imaging method. This method however is a destructive method, requiring the TBC coat to be separated from the substrate for the measurement. Alternatively, laser flash test may also be conducted on specially-prepared TBC specimens, including the top coat, bond coat, and substrate. TBC conductivity is then determined based on multilayer material analysis for the TBC system [4]. This method still requires two-sided access of the specimen, so it cannot be used to analyze TBCs coated on real components with variable substrate thickness.

In this study, one-sided pulsed thermal imaging is investigated for determination of thermal conductivity and heat capacity (the product of density and specific heat) of TBC coatings. Although thermal imaging has already been widely applied for nondestructive evaluation (NDE) of TBCs, such as for detection of delamination [5] and estimation of TBC thickness and conductivity [6.7], the analysis methods are generally empirical or semi-empirical because a fundamental understanding of the thermal response to various TBC parameters has not been established. This paper describes a new multilayer thermal-modeling method for TBC property characterization and imaging [8]. The accuracy for predicted TBC thermal properties is investigated for typical TBC systems and different sample preparation conditions. Results from experimental and theoretical analyses are presented and discussed.

PULSED THERMAL IMAGING ANALYSIS FOR MULTILAYER TBC MATERIALS

Pulsed thermal imaging is based on monitoring the temperature decay on a specimen surface after it is applied with a pulsed thermal energy that is gradually transferred inside the specimen. A schematic one-sided pulsed-thermal-imaging setup for testing a 3-layer material system is illustrated in Fig. 1. The premise is that the heat transfer from the surface (or surface temperature/time response) is affected by internal material structures and properties and the presence of flaws such as cracks [9,10]. By analyzing the surface temperature/time response, the material property and depth of various subsurface layers under the surface can be determined.

The important TBC parameters to be determined by thermal imaging include the thickness L, thermal conductivity k, and heat capacity ρc (where ρ is density and c is specific heat) of the top ceramic TBC layer. These three TBC parameters, however, may not be independent so may not be determined individually from the thermal imaging test. This problem is inherent to many thermal measurement methods. For example, the well-known laser flash method can only determine one single parameter from the same set of three parameters in a single-layer material. Normally, the thickness and the heat capacity of a sample have to be measured separately, so its thermal diffusivity and conductivity can be determined from the laser flash test.

Optical properties of the TBC layer, which is optically translucent, will also affect the data obtained from pulsed thermal imaging because both thermal excitation and sensing are based on optical characteristics of the material. To avoid the optical translucency issue and the requirement to determine optical properties that are normally not important for TBC performance, TBCs are usually coated by a thin graphite-based black paint when conducting thermal imaging tests. The problems associated to these paints are discussed and analyzed later.

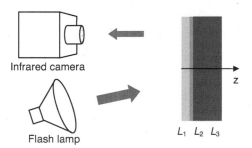

Fig. 1.   Schematics of pulsed thermal imaging of a 3-layer material system.

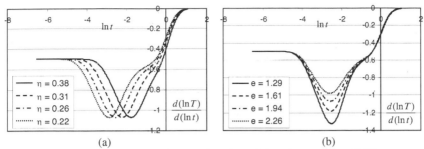

Fig. 2.   Surface temperature decay slope as function of time for two-layer TBC systems with variation of TBC (a) parameter η (s$^{1/2}$) and (b) effusivity e (kJ/m^2-K-s$^{1/2}$) while all other parameters are kept at constant.

A theoretical analysis [11,12] was conducted to identify the number of independent TBC parameters that can be determined from a one-sided thermal imaging test. For a two-layer TBC system where the bond coat is considered as part of the substrate and the TBC top coat is opaque (TBC surface is covered by a black paint), the analytical solution for the surface temperature as a function of time, which is measured in thermal imaging, contains only two independent parameters for each layer: the thermal effusivity e ($= (k\rho c)^{1/2}$) and the parameter $\eta = L/\alpha^{1/2}$ where α ($= k/\rho c$) is the thermal diffusivity. The dependency of the solution to the two TBC parameters becomes clear when examining the surface temperature decay slope, $d(\ln T)/d(\ln t)$, as function of time. Figure 2 shows the front surface temperature slope $d(\ln T)/d(\ln t)$ for TBCs with a varying parameter η (Fig. 2a) or effusivity e (Fig. 2b) while keeping the other parameter constant. The most visible characteristic of the temperature slope data is the large negative peak. This negative slope peak is the result of an increased heat conduction rate when the absorbed surface heat from the flash lamp is transferred from the topcoat into the substrate that has a higher thermal effusivity than the topcoat. Both the position (or transition time) and the magnitude of the peak change with the TBC thermal properties. In Fig. 2a, it is seen that the transition time for the slope change from -0.5 to -1.07 (the negative peak) is related only to the parameter η. Therefore, it is concluded that the parameter η is an independent parameter that determines the transition time for temperature-slope change, as predicted from theoretical analysis. On the other hand, the TBC effusivity e affects only the maximum slope value (see Fig. 2b). This result further confirms that among the three TBC parameters, thickness L and two thermal properties, only two of them can be determined uniquely from thermal imaging data.

A multilayer thermal modeling method was developed to facilitate thermal imaging analysis of multilayer materials such as TBCs [8]. In this method, a TBC is modeled by a multilayer material system and the 1D heat-transfer equation governing the pulsed thermal-imaging process is solved by numerical simulation. The numerical formulation may also incorporate other factors related to experiment or sample conditions; e.g., the finite flash duration [13] of the flash lamps and the TBC translucency which causes volume heat absorption effect. The numerical solutions (of surface temperature decay) are then fitted with the experimental data at each pixel by least-square minimization to determine unknown parameters in the multilayer material system. Multiple parameters in one or several layers can be determined simultaneously.

For analysis of opaque TBCs, two TBC properties are directly predicted: the thermal effusivity e and the parameter η. With a known TBC thickness L, TBC thermal conductivity k and heat capacity ρc are determined from: $k = Le/\eta$, $\rho c = e\eta/L$. This data fitting process is automated for all pixels within the thermal images and the final results are presented as images of the predicted TBC parameters.

THERMAL PROPERTY MEASUREMENT FOR TBCS

Although the fundamental theory for thermal imaging of multilayer TBC materials is rigorous and simple, the prediction accuracy can be affected by many factors from experimental and sample condition variations. The experimental factors may come from various sources. First, the accuracy of the measured surface temperature needs to be addressed. It is understood that infrared thermal imaging does not measure temperature, but the intensity of infrared radiation received by the detectors within the infrared camera. The relationship between the detected radiation intensity and the equivalent blackbody temperature can be calibrated, but the calibration may vary with time. The measured blackbody temperature in general is not equal to the surface temperature; corrections have to be made for the surface emissivity and the presence of any significant reflecting heat sources. One such source is the flash lamp. Heat reflection from flash lamps can be significant especially during the flashing period, and its effect cannot be corrected. To address these issues, a dynamic calibration procedure was developed to determine the correct surface temperature and an infrared filter was used to completely eliminate the flash-lamp infrared radiation [14].

At least two factors from TBC samples may affect the measurement accuracy: the condition of the black paint normally applied on TBC surface for thermal or laser flash test and the TBC surface roughness. The black paint increases the surface emissivity so improves heat absorption efficiency and infrared radiation intensity (therefore detection sensitivity). It also prevents the flash energy from penetrating inside the TBC layer. Although the black paint has been used in almost all quantitative TBC characterizations by laser flash and thermal imaging, its effect to measurement accuracy has not been studied. As to be discussed in the following, the condition of the paint may significantly affect the measurement results for thin TBC coatings. On the other hand, the roughness of the TBC surface has complex effect to the surface temperature transient; it is not explored under this study.

The measurement accuracy was investigated using as-processed APS (air plasma sprayed) TBC specimens with known TBC thickness. Measurement data for a thick TBC specimen (sample curtsey of Dr. Y. Tan of Stony Brook University, NY) were compared with those typical for such materials obtained from laser flash tests. The top ceramic coating is 0.86 mm thick and the substrate is a stainless steel material with a thickness of 10 mm. The TBC surface was coated by a black paint for the thermal imaging test. The predicted average TBC conductivity was 0.93 W/m-K, which is consistent with the measured values of 1.0 ± 0.2 W/m-K from laser flash tests. The predicted average TBC heat capacity was 2.19 J/cm^3-K, which is also in good agreement with typical value of 2.0 J/cm^3-K for this TBC. The minor differences are probably due to the material variation in individual samples and small variations in the TBC thickness. To verify the prediction accuracy, Fig. 3 compares the experimental data and the predicted theoretical data for a typical pixel within the TBC sample; both the location and the magnitude of the temperature-slope peak are matched accurately as shown in Fig. 3b. The slight discrepancies in early time period ($t < 0.02$s) was likely due to the surface roughness and effects related to the black paint.

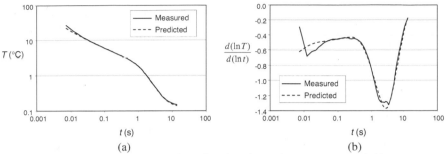

Fig. 3. Comparison of measured and predicted surface (a) temperature and (b) temperature-derivative data for a typical pixel in the 0.86-mm-thick TBC sample.

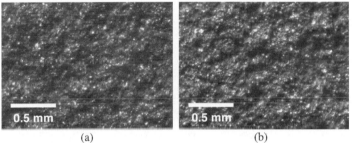

Fig. 4. Photomicrographs of APS TBC surface coated with (a) Paint #1 and (b) Paint #2.

Black paints have been used in most of the thermal imaging and laser flash tests for TBCs. These paints consist of micron-sized graphite particles and some binder materials. Depending on the composition, they may have different thermal properties (conductivity and heat capacity). When applied on TBC surface, a sufficient thickness of the paint layer is required to prevent the light penetration through the paint. As a result, the black paint may have to be considered as an additional layer of material on the TBC surface when its thermal effect cannot be neglected. This effect comes from the combination of the thickness and the thermal properties of the paint, as well as a possible infiltration of the paint inside the pores of the TBC coating. To investigate the effect of the paint, two graphite-based paints, both are normally used as dry-film lubricant, were tested on TBC samples of various thicknesses. Paint #1 is a heavy-duty lubricant and Paint #2 is a long-wearing lubricant with fine graphite particles. Because they are normally sprayed manually on TBC surfaces, their thicknesses cannot be controlled and are therefore unknown. Intuitively, it is believed that the film thickness of Paint #1 is usually thicker than that of Paint #2. Figure 4 shows micrographs of the painted surfaces using the two paints; no apparent difference is observed.

Paint #1 was used on the 0.86-mm-thick TBC sample shown in Fig. 3 and in previous studies [12]. For TBC coatings with thickness ≥0.3 mm, it was found that measured thermal properties are essentially the same for TBC samples coated with both paints. However, when an additional layer of the same paint was applied, predicted thermal properties for the TBC sample coated with Paint #1 showed considerable increases, while those with Paint #2 did not change. Similarly, the predicted thermal properties for thinner TBCs were higher in Paint #1 coated samples. One example is shown in [12], where the predicted heat capacity was much higher than the expected value for a 127-μm-thick EBPVD TBC. To investigate the effect of the paints, a 152-μm-thick APS TBC sample was coated with each paint on half of its surface, and a thermal imaging test was conducted for this sample. The average temperature-slope curves in the two painted surface areas are shown in Fig. 5. A significant difference is observed for both position and magnitude of the negative peak: it occurs at an earlier time and has a low magnitude in Paint #1 area. Consequently, the predicted thermal properties within the Paint #1 surface are much higher, $k = 0.93$ W/m-K and $\rho c = 3.27$ J/cm^3-K, compared with those in Paint #2 surface, $k = 0.80$ W/m-K and $\rho c = 3.06$ J/cm^3-K, although the fitting for the peak in both surface areas is very good (see Fig. 6). The small difference in early times (before ~0.02s) may be related to the translucency of the paint and the surface roughness; this is not further studied here. This result demonstrates the importance in selection and application of the black paint for thermal imaging and likely laser flash tests for TBC samples, and further studies should be considered to understand the mechanisms.

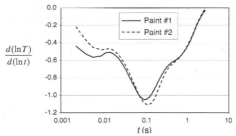

Fig. 5.   Temperature-derivative data on surface areas coated with Paint #1 and #2.

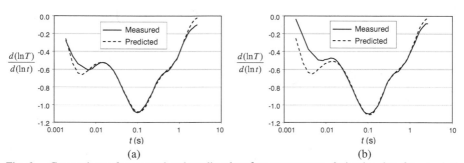

Fig. 6.   Comparison of measured and predicted surface temperature-derivative data for a typical pixel in (a) Paint #1 and (b) Paint #2 coated TBC surface.

CONCLUSION

A multilayer thermal modeling method was developed to analyze the surface temperature response for TBCs under one-sided pulsed thermal imaging test condition. The model may be used to predict two independent TBC topcoat parameters: the thermal effusivity e and the parameter $\eta - L^2/\alpha$. With a known TBC thickness L, TBC thermal conductivity k and heat capacity ρc can be determined from: $k = Le/\eta$, $\rho c = e\eta/L$. One significant advantage of thermal imaging is that it can simultaneously determine two thermal properties, k and ρc, while each would require a separate test by conventional methods. The multilayer thermal modeling method was used to measure the thermal properties of a 0.86-mm-thick APS TBC sample, and both predicted conductivity and heat capacity are in good agreement with known data for this material. It was found that the application of black paint on TBC surface may affect the measurement accuracy. Two paints were investigated, and the results showed that one paint consistently induces higher predicted values for thin TBCs. With further investigations on the mechanisms of the black-paint effect as well as other effects such as the surface roughness, thermal imaging will be further established for accurate property measurement as well as NDE characterization for TBCs.

ACKNOWLEDGMENT

The author thanks Dr. Y. Tan of Stony Brook University for providing test specimens for this study. This work was sponsored by the U.S. Department of Energy, Office of Fossil Energy, Advanced Research and Technology Development/Materials Program, and by the Heavy Vehicle Propulsion Materials Program, DOE Office of FreedomCAR and Vehicle Technology Program, under contract DE-AC05-00OR22725 with UT-Battelle, LLC.

REFERENCES

1. US National Research Council, National Materials Advisory Board, "Coatings for High Temperature Structural Materials," National Academy Press, Washington, DC, 1996.
2. A. Feuerstein and A. Bolcavage, "Thermal Conductivity of Plasma and EBPVD Thermal Barrier Coatings," Proc. 3rd Int. Surface Engineering Conf., pp. 291-298, 2004.
3. H. Wang and R.B. Dinwiddie, "Reliability of laser flash thermal diffusivity measurements of the thermal barrier coatings," J. Thermal Spray Techno., Vol. 9, pp. 210-214, 2000.
4. B.K. Jang, M. Yoshiya, N. Yamaguchi, and H. Matsubara, "Evaluation of Thermal Conductivity of Zirconia Coating Layers Deposited by EB-PVD," J. Mater. Sci., Vol. 39, pp. 1823-1825, 2004.
5. X. Chen, G. Newaz, and X. Han, "Damage Assessment in Thermal Barrier Coatings Using Thermal Wave Image Technique", Proc. 2001 ASME Int. Mech. Eng. Congress Expo., Nov. 11-16, 2001, New York, NY, paper no. IMECE2001/AD-25323, 2001.
6. H. I. Ringermacher "Coating Thickness and Thermal Conductivity Evaluation Using Flash IR Imaging," presented in Review of Progress in QNDE, Golden, CO, July 25-30, 2004.
7. S. M. Shepard, Y. L. Hou, J. R. Lhota, D. Wang, and T. Ahmed, "Thermographic Measurement of Thermal Barrier Coating Thickness," in Proc. SPIE, Vol. 5782, Thermosense XXVII, 2005, pp. 407-410, 2005.
8. J.G. Sun, "Thermal Imaging Characterization of Thermal Barrier Coatings," in Ceramic Eng. Sci. Proc., eds. J. Salem and D. Zhu, Vol.28, no. 3, pp. 53-60, 2007.
9. J. G. Sun, "Analysis of Pulsed Thermography Methods for Defect Depth Prediction," J. Heat Transfer, Vol. 128, pp. 329-338, 2006.

10. J. G. Sun, "Evaluation of Ceramic Matrix Composites by Thermal Diffusivity Imaging," Int. J. Appl. Ceram. Technol., Vol. 4, pp. 75-87, 2007.
11. J.G. Sun, "Thermal Imaging Analysis of Thermal Barrier Coatings," in Review of Quantitative Nondestructive Evaluation, eds. D.O. Thompson and D.E. Chimenti, Vol. 28, pp. 495-502, 2008.
12. J.G. Sun, "Measurement of Thermal Barrier Coating Conductivity by Thermal Imaging Method," in Ceramic Eng. Sci. Proc., eds. D. Singh and J. Salem, Vol. 30, no. 3, pp. 97-104, 2009.
13. J. G. Sun and J. Benz, "Flash Duration Effect in One-Sided Thermal Imaging," in Review of Progress in Quantitative Nondestructive Evaluation, eds. D.O. Thompson and D.E. Chimenti, Vol. 24, pp. 650-654, 2004.
14. J.G. Sun, "Optical Filter for Flash Lamps in Pulsed Thermal Imaging," U.S. Patent No. 7,538,938 issued May 26, 2009.

RESEARCH ON MoSi$_2$–BASED OXIDATION PROTECTIVE COATING FOR SiC-COATED CARBON/CARBON COMPOSITES PREPARED BY SUPERSONIC PLASMA SPRAYING

Cao Wei, Li Hejun ,Wu Heng, Xue Hui, Fu Qiangang, Ma Chao, Wang Yongjie, Wei Jianfeng
Carbon/Carbon Composites Research Center, Northwestern Polytechnical University
Xi'an, Shannxi, P. R. China

ABSTRACT

In order to improve the oxidation protective ability of the SiC-coated (C/C) composites, a MoSi$_2$-based coating was prepared on them by a novel supersonic plasma spraying method. Microstructures of the coating were analyzed by XRD and SEM. XRD result indicates the MoSi$_2$-based coating is primarily composed of MoSi$_2$, Mo$_5$Si$_3$ and SiO$_2$. SEM analyses show that the MoSi$_2$-based coating is dense and crack-free, and has a good interface bonding with the SiC-coated C/C composites. Oxidation test shows that the coating has excellent oxidation and thermal shock resistance, and can effectively protect C/C composites from oxidation at 1673K in air for more than 500h and endure the thermal cycle between 1673K and room temperature for 15 times. The outstanding oxidation protective ability mainly attributes to the formation of SiO$_2$ film on the surface of the coating.

INTRODUCTION

Carbon/carbon (C/C) composites are one of the most promising thermal structure composites for a number of applications including engineering materials or advanced vehicles, leading edges, rocket nozzles, disk brake and propulsion systems. However, these composites will be oxidized during their exposure to an oxidizing atmosphere above 723K. The oxidation will result in an obvious decrease of their mechanical properties hence limit their applications as high temperature structural materials dramatically.

Oxidation resistant coating is considered as an extremely effective approach to protect C/C composites from oxidation at high temperatures. SiC ceramic has been used widely as the coating for protecting C/C composites at high temperature owing to its excellent anti-oxidation property and good compatibility with C/C composites. It is often considered as bonding or buffer layer in multi-layer coating. In addition, MoSi$_2$ has high melting point, excellent high-temperature oxidation and corrosion resistance due to the formation of vitreous SiO$_2$ glass film, which makes it an attractive material as outer layer. Supersonic plasma spraying technique is a new coating method developed in recent years, and has been used widely due to its ability to prepare coating. In supersonic plasma spraying, the velocities of the plasma jet and particles of 2300-2500m/s and 500-600m/s can be achieved at the normal distance of 100 mm from the nozzle exist, the temperature of plasma jet near nozzle exist could be up to 10000℃, at which the spraying powders are heated to a molten state. the molten powders is propelled in plasma jet at high velocities to impact the substrate, meantime, they become flattened and solidified very soon and eventually form a stack of lamellae of coating with compact microstructure,

Corresponding author.
E-mail address: lihejun@nwpu.edu.cn (H.J. Li)

high hardness, low porosity and excellent bonding strength. Up to now, no literature has been published on the preparation of MoSi$_2$-based coating for the oxidation protection of C/C composites by supersonic plasma spraying technique.

In the present work, in order to improve the oxidation resistance of C/C composites in air at 1673K, MoSi$_2$-based coating was prepared on the surface of the SiC-coated C/C composites by supersonic plasma spraying. The microstructures and oxidation protective ability of the coating were investigated.

EXPERIMENTAL

Small specimens with the dimensions of $\Phi25\times5$ mm were cut from a bulk of 2 D C/C composites with a density of 1.70 g/cm^3. Preforms used in this study were carbon fibers needled felts and they were densified by thermal gradient chemical vapor infiltration. After being polished with 100, 300 and 500 grit SiC paper, the specimens were ultrasonically cleaned with acetone and dried at 373 K for 1 h. The SiC coating was prepared by a pack cementation process with Si, C, and B$_2$O$_3$ powder. Powder compositions were as follows70-80% Si, 10-20% graphite and 5-10% B$_2$O$_3$. The role of B$_2$O$_3$ added in the pack powders was to increase the rate of diffusing reaction. All the above powders, with the granularities of 325 meshes, were analytical grade and were mixed in a ball mill for up to 2 h. The pack mixtures and C/C composites were put into a graphite crucible. The crucible was placed into an electronic furnace, and then was heated to 1973-2173K and held at that temperature for 2 h under slight argon flow to form SiC coating. The MoSi$_2$-based coating for SiC-coated C/C composites was prepared by supersonic plasma spraying. The spraying system mainly consists of jet gun, powder feeder, gas supply, water cooling circulator, control unit and power supply. Diagram of supersonic plasma spraying equipment is showed as Fig.1. A commercially available MoSi$_2$ powder was used as spraying materials, with the purity of 99.9wt.% and the particle size in the range 5-10μm. Details of the spraying parameters are listed in Table 1.

The oxidation test was carried out at 1673K in air in an electrical furnace. The sample was weighed at room temperature by electronic balance with a sensitivity of ±0.1mg during oxidation at 1673K. The morphology and crystalline structure of the coating were analyzed by scanning electron microscopy (SEM) and Rigaku D/max-3C X-ray diffraction (XRD).

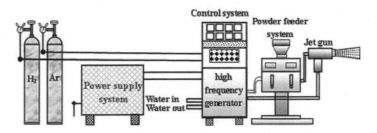

Fig.1 Diagram of supersonic plasma spraying equipment

Table 1. Details of the spraying parameters for the MoSi$_2$-based coating.

Spraying voltage	370-400V
Spraying current	115-140A
Primary gas (Ar) flow rate	3.6-4.4m^3/h
Secondary gas (H$_2$) flow rate	1.0-1.5m^3/h
Carrier gas (Ar) flow rate	0.016-0.024 m^3/h
Powder feed rate	about 20g/min
Spraying distance	100 mm
Nozzle diameter	5.5mm

RESULTS AND DISCUSSION

Microstructure of the coating

Fig. 2 shows the X-ray diffraction patterns of the surface of the SiC coating prepared by pack cementation. It reveals the inner coating consists of SiC and Si. The residual Si could relax the mismatch of thermal expansion coefficient between SiC coating and C/C composites. Meanwhile, it is helpful to improve bonding ability between the inner SiC coating and the outer MoSi$_2$-based coating.

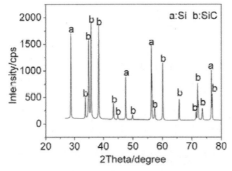

Fig.2 X-ray patterns on the surface of the SiC coating

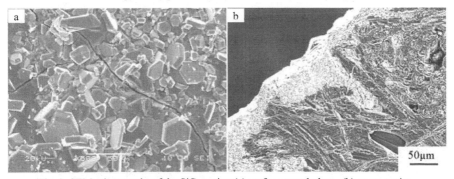

Fig. 3. SEM micrographs of the SiC coating (a) surface morphology, (b) cross-section.

Fig.3 shows the surface and cross-section SEM micrographs of the SiC coating. It reveals that there are some cracks in the SiC coating and the thickness is about 50μm. At high temperatures, Si would melt and could infiltrate into the C/C composites deeply through some holes or interfaces between carbon fibers and carbon matrix. The interleaving interface structure is advantageous to improve the bonding strength between the SiC coating and the C/C substrate.

Fig.4 shows the surface XRD of the MoSi$_2$-based coating. It can be seen that the coating consists of MoSi$_2$, SiO$_2$ and Mo$_5$Si$_3$. MoSi$_2$ comes from the original powder composition. During the process of spraying, the plasma arc temperature could reach around 10000K, so MoSi$_2$ powders could melt quickly and react with oxygen to form SiO$_2$, Mo$_5$Si$_3$ and MoO$_3$ according to Eqs (1) and (2). Even a small amount of Mo$_5$Si$_3$ could be further oxidized to produce SiO$_2$ and MoO$_3$ according to Eqs (3). Because of the volatilization of MoO$_3$ at high temperature, there are only MoSi$_2$, Mo$_5$Si$_3$, and SiO$_2$ in the coating. SiO$_2$ with low viscosity and good fluidity could efficiently fill microcracks and pinholes in coating at high temperature. Moreover, Mo$_5$Si$_3$ could improve the stability of SiO$_2$ glass and reduce its volatilization. Therefore the double-phase structure of SiO$_2$ and Mo$_5$Si$_3$ is advantageous to improve the oxidation resistance of the coating, which indicates that the partial oxidation of MoSi$_2$ powder has no negative effect on the oxidation resistance of the coating.

$$5MoSi_2 + 7O_2 = Mo_5Si_3 + 7SiO_2 \qquad (1)$$
$$2MoSi_2 + 7O_2 = 4SiO_2 + 2MoO_3 \qquad (2)$$
$$2Mo_5Si_3 + 21O_2 = 10MoO_3 + 6SiO_2 \qquad (3)$$

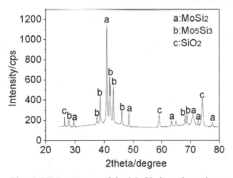

Fig. 4. XRD pattern of the MoSi$_2$-based coating

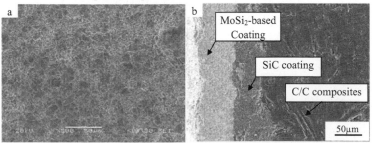

Fig. 5. SEM micrographs of the MoSi$_2$-based coating (a) surface, (b) cross-section.

Fig.5a shows the surface morphology of the MoSi$_2$-based coating. The coating is dense, on which only some microcracks and a few pinholes can be observed. The cross-section image (Fig.5b) shows that the MoSi$_2$–based coating is about 50μm in thickness without penetrating crack or hole. Moreover, no obvious crack or hole at the interface between the inner SiC coating and the outer MoSi$_2$-based coating is observed, which indicates a good bonding.

Oxidation protective ability of the coating

The isothermal oxidation curves of MoSi$_2$-based coating for SiC coated C/C composites in air at 1673K are illustrated in Fig.6. It can be seen that the as-prepared coating exhibits excellent oxidation protective ability. After oxidation in 1673K in air for 500h and thermal cycle between 1673K and room temperature for 15 times, the mass loss is negligible. However, the mass loss of the SiC-coated composites reach 10Wt.% after oxidation only for 15h, the oxidation weight loss is mainly attributable to the SiC coating with many large cracks which are difficult to self-healing at high temperature. In order to reveal the oxidation mechanism of the MoSi$_2$-based coating for SiC coated C/C composites, the oxidation behavior could be divided into three processes from the oxidation curves (Fig.6). In the initial stage of oxidation (0-5h), the specimens were exposed to the atmosphere. MoSi$_2$ and Mo$_5$Si$_3$ in the MoSi$_2$-based coating were oxidized to form MoO$_3$ and SiO$_2$ as Eqs. (2) and (3) due to high oxygen pressure. The volatilization of MoO$_3$ covered mass gain coming from the formation of SiO$_2$ glass, which resulted in the quick mass loss of the specimens. In the second period (5-410 h), SiO$_2$ film was generated well on the MoSi$_2$-based coating, which reduced the partial pressure of oxygen and made MoSi$_2$ below SiO$_2$ film react with oxygen incompletely as Eqs.(1). Therefore the coated specimen exhibits mass gain. Between 410h and 500h, The specimen loses mass slowly, which might be resulted from the formation of some microcracks in the SiO$_2$ glass film.

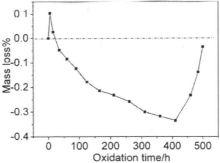

Fig. 6. Isothermal oxidation curves of MoSi$_2$-based coatings for SiC coated C/C composites
in air at 1673 K

Fig. 7. shows the XRD pattern of the coating after oxidation at 1673K for 500h. There are two kinds of crystal structures in the surface of the coating, including SiO$_2$ and Mo$_5$Si$_3$. Compared with the X-ray patterns of the coating before oxidation (Fig. 4), the diffraction peak intensity of SiO$_2$ increases and MoSi$_2$ phase disappears. During the oxidation test, the MoSi$_2$-based coating was consumed due to the reaction between MoSi$_2$ and oxygen in air according to Eqs (1) and (2), resulting in the disappearance of the MoSi₂ phase on the surface of the coating.

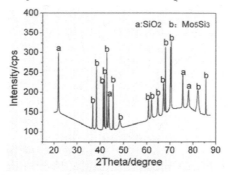

Fig. 7. XRD pattern of the MoSi₂-based coating after oxidation at 1673K for 500 h.

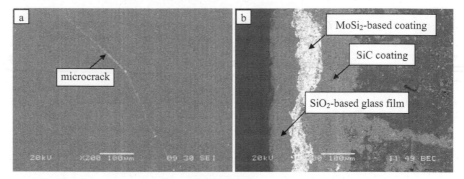

Fig. 8. SEM images of the MoSi₂-based coating after oxidation for 500 h (a) surface, (b) cross-section.

The surface and cross-section images of the MoSi₂-based coating for SiC-coated C/C composites after oxidation at 1673K for 500h are shown in Fig.6. From Fig.6 a, it can be seen that the surface of the coating has changed into smooth glass due to the formation of SiO₂-based film by the oxidation of MoSi₂ and Mo₅Si₃ in the coating at 1673K in air, which can efficiently prevent oxygen from diffusing into the C/C substrate. Because SiO₂ with low vapor pressure hardly volatilizes at 1673K, the SiO₂-based film is gradually thickened. During the quick cooling from 1673K to room temperature for weighing up during the isothermal oxidation test, large thermal stress was produced in the SiO₂-based film, which leads to the formation of microcracks. These microcracks would self-seal when the coated C/C sample was heated to 1673K again. However, MoSi₂-based coating would be oxidized to form MoO₃ and SiO₂ as Eqs. (2) and (3) by oxygen diffusing through the microcracks in the coating below the crack sealing temperature, which is the main reason for the weight loss of the specimens after oxidation for 410 h. From the cross-section image of the coated sample after oxidation, a three-layer coating can be clearly seen without cracks or big holes, as presented in Fig. 8 (b). Compared with the cross-section image of the coating before oxidation, a SiO₂-based glass film is formed on the surface of the MoSi$_2$-based coating, owing to the oxidation of MoSi$_2$-based coating. The outstanding oxidation protective ability mainly attributes to the formation of SiO$_2$-based glass film on the coating surface. Moreover, the MoSi$_2$-based coating exhibits excellent thermal shock resistance because no cracks or debonding is found after thermal cycles between 1673 K and room temperature for 15 times.

CONCLUSION

A dense MoSi$_2$-based coating has been prepared on SiC-coated C/C composites by supersonic plasma spraying, which primarily consists of MoSi$_2$, Mo$_5$Si$_3$ and SiO$_2$, The coating possesses excellent oxidation protective ability and good thermal shock resistance. It could effectively protect SiC-coated C/C composites from oxidation for more than 500 h at 1673 K in air, and endure the thermal cycle between 1673K and room temperature for 15 times. The outstanding oxidation protective ability mainly attributes to the formation of SiO$_2$-based glass film on the surface of MoSi$_2$-based coating.

ACKNOWLEDGMENTS

This work has been supported by the National Natural Science Foundation of China under Grant No.90716024 and No.50802075, and the "111" Project under Grant No.08040 and the Research Fund of State Key Laboratory of Solidification Processing (NWPU), China (Grant No.KP200913).

REFERENCES

[1] J. D. Buckley, Carbon-carbon, an overview, Ceram Bull., 67, 364-368(1988).

[2] J.R. Strife, J.E. Sheehan, Ceramic coatings for carbon–carbon metallic dispersions, Ceram. Bull., 67, 369-374. (1988).

[3] F. Smeacetto, M. Salvo, M. Ferraris, Oxidation protection multilayer coatings for carbon/carbon composites, Carbon, 40, 583-587(2002).

[4] Q. G. Fu, H. J. Li, X. H. Shi, K. Z. Li, G.D. Sun, Silicon carbide coating to protect carbon/carbon composites against oxidation. Scripta Mater., 52, 923-927(2005).

[5] J. Cook, A. Khan, E. Lee, R. Mahapatra, Oxidation of MoSi$_2$-based composites, Mater. Sci. and Eng: A, 155, 183-198(1992).

[6] J. H. Schneibel, J. A. Sekharb, Microstructure and properties of MoSi$_2$–Mo-B and MoSi$_2$–Mo$_5$Si$_3$ molybdenum silicides Mater. Sci. and Eng. A, 340, 204-211(2003).

[7] Z.H. Han, B. S. Xu, H. J. Wang, S.K. Zhou,. Microstructures, mechanical properties, and tribological behaviors of Cr–Al–N, Cr–Si–N, and Cr–Al–Si–N coatings by a hybrid coating system. Surf. Coat. Technol., 201,5253-5257(2007).

[8] X. C. Zhang, B. S. Xu, Y. X. Wu, F. Z. Xuan, S. T. Tu, Porosity, mechanical properties, residual stresses of supersonic plasma-sprayed Ni-based alloy coatings prepared at different powder feed rates, Appl. Surf. Sci. ,254, 3879-3889(2008).

[9] S. Vaidyaraman, W. J. Lackey, P. K. Agrawal, T. L. Starr, 1-D model for forced flow-thermal gradient chemical vapor infiltration process for carbon/carbon composites, Carbon, 34, 1123-1133(1996).

[10] Y. L. Zhang, H. J. Li, Q.G. Fu, K.Z. Li, A Si–Mo oxidation protective coating for C/SiC coated carbon/carbon composites, Carbon; 45, 1130-1133(2008).

Advanced Coatings for Wear and Corrosion Applications

LAFAD-ASSISTED PLASMA SURFACE ENGINEERING PROCESSES FOR WEAR AND
CORROSION PROTECTION: A REVIEW

V. Gorokhovsky
Southwest Research Institute®
San Antonio, Texas, United States

ABSTRACT
    The large area filtered arc deposition (LAFAD) process is characterized for deposition of
various metal, ceramic and cermet coatings having different architectures: from monolithic ceramic to
nano-microlaminated, nanolayered and sub-stoichiometric coatings. Duplex processes, consisting of a
bottom ionitrided segment followed by coating layers, have been tested for application in forming
tools, dies and molds. The barrier properties of LAFAD coatings are mostly due to its nearly
defect-free morphology and smoothness as it was demonstrated in die casting die applications and in
the application of protective coatings for solid oxide fuel cell (SOFC) metallic interconnects, operating
in oxidizing environment at 800°C. Ultra-thick ceramic and cermet coatings with thicknesses ranging
from 30 to 120 µm have been deposited by one unidirectional dual arc LAFAD vapor plasma source
onto substrates installed on a rotating turntable 0.5 m in diameter in an industrial coating system with
deposition rates exceeding 5 µm/hr. The operating pressure range of the LAFAD process allows it to
be used in plasma-assisted hybrid technologies with conventional magnetron sputtering and electron
beam evaporation processes. The industrial applications of LAFAD wear and corrosion resistant
coatings; their performance in different applications; and a commercialization strategy for LAFAD
technology will be discussed.

INTRODUCTION
    The contemporary surface engineering technology has reached the ability to grow layered
structures in a controlled way, leading ultimately to a new generation of electronic and optical devices,
cutting and forming tools operating at higher speeds, protecting the turbomachinery components
against erosion and corrosion in a harsh environment, widening the horizons of new energy related
applications from fuel cells to power stations, to automotive and aircraft components operating at high
contact stress, among many others. In order to meet requirements for coatings operating in extreme
conditions such as high contact stress, hostile environments, high temperatures and vibration, coating
properties such as adhesion and cohesion toughness, fracture resistance, corrosion and high
temperature oxidation resistance must be improved compared to the present state-of-the-art coatings.
    It is well established that assistance of the coating deposition process with bombardment by
energetic particles, especially energetic metal ions, can dramatically improve coatings by densifying the
depositing materials, reducing the grain size and reducing or completely eliminating the growth
defects[4,5,27,58]. It is also a mechanism for improving coating adhesion and cohesion toughness by mixing
the substrate atoms with the atoms of the depositing coating, or mixing the atoms of the neighbor
sublayers in laminated coating architectures by ion-bombardment assisted deposition processes[1,2,4-9,26,27,57,58]. In coating processes assisted with ion bombardment, the surface layer is affected by a high rate
of bombardment by energetic ions, which affects the mobility of depositing metal vapor atoms and in
many cases creates metastable structures with unique functional properties[1,4,26,27,57]. In addition, the ion
bombardment of the coating surface influences gas adsorption behavior by increasing the sticking
coefficient of gases such as nitrogen and changing the nature of adsorption sites from lower energy
physisorption sites to higher energy chemisorption sites. This approach is especially productive in
deposition of nanostructured nanocomposite coatings with ultra-fine or glass-like amorphous structure[5,9].
The ion-to-atoms arrival ratio represents one of the most important characteristics of such
processes[4,8,27,58]. It is essential that this ratio is calculated at the atom deposition spot. If ion bombardment

is provided in a different location of the coating chamber or even at the distant spot and at different times from the atom condensation site, this cannot be considered as ion assisted deposition but rather as a post deposition ion treatment. The reason for this is that the time during which the adatom is fully accommodated to the surface lattice is less than 10^{-13} s (the accommodation time), which requires almost immediate ion- bombardment assistance at the atom condensation spot.

IONIZED VAPOR DEPOSITION TECHNOLOGIES

Conventional physical vapor deposition (PVD) sources, such as electron beam evaporator (EBPVD) and direct current magnetron sputtering (DCMS) sources can provide high deposition rates, but low energy of the metal vapor atoms results in low density, poor adhesion, poor structure and morphology of the coatings with high concentration of growth defects[1,2,4,26,27,58]. In the ion-beam assisted deposition (IBAD) process the flow of condensing metal vapor atoms is assisting concurrently by gaseous ion beam created by a separate ion beam source, such as Kauffman type gridded source or end-Hall griddles source as shown schematically in Figure 1. IBAD processes can provide gaseous ion bombardment assistance; however are not capable of providing metal ion bombardment assistance and are not capable of assisting in the deposition of conformal coatings on components with complex shapes[8]. The advanced plasma enhanced magnetron sputtering (PEMS) process, schematically shown in Figure 2 uses a distant thermionic or hollow cathode for ionizing gaseous environment in magnetron sputtering deposition process. In this process the deposition of sputtered metal atoms is assisted by concurrent ion bombardment from ionized gaseous plasma environment but the metal sputtering flow is mostly non-ionized as in the conventional DCMS process[9,56]. The recently developed high power impulse magnetron sputtering (HIPIMS) process uses high power pulses applied to the magnetron target which results in increased electron emission, and consequently increases the ionization rate of the metal sputtering flow[7]. This process has

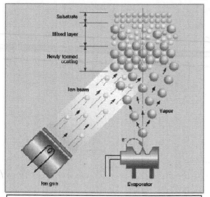

Figure 1. Schematic illustration of IBAD coating process.

demonstrated that increased ionization rate of the metal sputtering atoms results in improved coating density and adhesion, and ultimately leads to improved coating properties in the deposition of nitride wear resistant coatings for cutting tools. The limitation of the HIPIMS process is that improved ionization is achieved only during the short pulse time. Since pulse parameters are coupled with magnetron sputtering process parameters in HIPIMS technology, it can adversely affect the sputtering rate which is found three to five times lower than that of the conventional DCMS process[7].

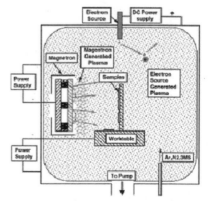

Figure 2. Schematic of plasma enhanced magnetron sputtering (PEMS).

LAFAD AND HYBRID SURFACE ENGINEERING SYSTEMS

The vacuum arc discharge plasma has been used successfully in the last two decades for the deposition of hard coatings for cutting tools and machine parts. In a direct cathodic arc deposition (DCAD) process the highly ionized metal vapor plasma is generated by direct cathodic arc sources (DCAS) in a form of a vapor plasma jet flowing from the cathodic arc spots, which transfers coating material from the target to the substrate surface. A significant disadvantage of this method is the formation of droplets, also known as macro-particles, in the cathodic arc jets, which limit the application of the process to surface coatings that do not require high precision or surface finish. These particles also deleteriously influence critical properties of the coatings as shown illustratively in

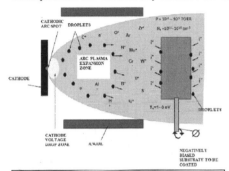

Figure 3. Schematic diagram of direct cathodic arc deposition (DCAD) process.

Figure 3. For instance in the case of TiN coating on cutting tools, the presence of titanium particles in the coating compromises the hardness and wear resistance of the coating[1,2,14,21].

One of the first macroparticle filters had a quarter-torus cylindrical electromagnetic plasma guide, and was based on the torus-type plasma traps design, which was developed for the controlled nuclear fusion apparatus such as the Stellarator and the Tokamak. The filter removed the macro-particles, but could operate only with small cathode targets and could not be scaled up due to the difficulty of scaling up the cylindrical magnetic coils[4,11,12,26,27,58]. The main obstacle to using conventional filtered cathodic arc deposition (FCAD) technology is the low productivity of this process, which restricts its usage to semiconductors, optical coatings and some ultra-thin hard coatings used in bio-medical and tribological applications[58]. On the other hand, LAFAD technology overcomes these limitations by providing a highly productive, robust, industry-friendly process which combines the high productivity rate of conventional DCAD and magnetron technologies, with the capability of generating a nearly 100% ionized metal-gaseous vapor plasma with large kinetic energy and with no macroparticles, droplets, multi-atom clusters and other contaminants[1,2,16,21,22,29,30,58]. The unidirectional dual arc LAFAD vapor plasma sources can be used as an alternative to conventional DCAD and magnetron based processes when the high productivity and uniformity needed for most industrial applications must be accompanied by the high ionization and high kinetic energy of atomically clean vapor plasma. Since the LAFAD plasma source operating pressure regimes are overlapping with most of the conventional vacuum vapor deposition technologies such as magnetron sputtering (MS), electron beam physical vapor deposition (EBPVD), thermal evaporation, and plasma assisted chemical vapor deposition (PACVD), it can be used in hybrid processes combining its high deposition and high ionization rates in conjunction with conventional PVD and low pressure PACVD processes as demonstrated from the referenced literature[1,2,15,29,43,47,48,58].

The hybrid filtered arc plasma source ion deposition (FAPSID) surface engineering system combining LAFAD process with ionitriding, ion implantation and conventional PVD and low pressure PACVD processes is shown schematically in Figure 4. The FAPSID system employs two unidirectional LAFAD dual filtered cathodic arc sources, two unbalanced magnetrons, as well as, electron beam physical vapor deposition (EBPVD) and thermal evaporation sources in one universal vacuum chamber layout[29,43,47,48]. The LAFAD plasma source magnetic deflecting system allows bending of the metal vapor plasma jets at 90° toward substrates to be coated, which yields 100% ionized metal vapor plasma flow at the LAFAD source exit and more than 50% ionized gaseous plasma in the coating chamber. Using a

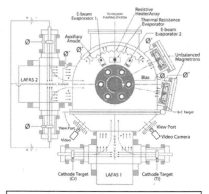

Figure 4. Schematic of filtered arc plasma source ion deposition (FAPSID) surface engineering system.

vertical rastering magnetic field allows obtaining uniform coating thickness distribution over large deposition areas suitable for industrial PVD coating systems[30]. When the magnetic deflecting subsystem is turned off the LAFAD source can be used in a gas ionization mode as a powerful electron emitter. In this mode the auxiliary arc discharge is ignited between the primary cathodes of the LAFAD source and an auxiliary anode located in the coating chamber as schematically illustrated in Figure 4. Multi-segment coating architecture utilizing segments deposited by LAFAD source followed by segments deposited by other PVD or low pressure PACVD sources, can be deposited by hybrid coating systems similar to that shown in Figure 4[29,43,47,48]. Alternatively, nanocomposite multi-phase, multielemental coatings can be deposited by concurrently using LAFAD with other sources such as unbalanced magnetron (UBM), EBPVD or low pressure PACVD sources[29,43,47,60,61,65,66].

LAFAD-ASSISTED PROCESSES AND APPLICATIONS

Various metal, ceramic and cermet coatings of different architectures where deposited by LAFAD process for different applications. Initially the efforts were directed toward development of hard ceramic coatings and superhard DLCs over large size machine parts or by processing of large number of small components in a fully loaded industrial scale coating chambers. The typical LAFAD process for deposition of monolithic (single layer) titanium nitride ceramic coatings includes pre-heating the components to be coated to 300-400°C depending on type of substrate material. The pre-heating stage is conducting in argon at the pressure ranging from 0.1 to 1 mtorr. The pre-heating stage is followed by ion cleaning stage. At this stage the dense argon auxiliary arc plasma discharge is established between cathode targets of the two primary direct cathodic arc sources (DCAS) of the LAFAD plasma source and auxiliary anode disposed in a coating chamber. After ion cleaning the high voltage metal ion etching stage is commenced for a duration of 2 to 5 min at argon pressure below 0.1 mtorr. During this stage the substrates to be coated are subjected to intense metal ion bombardment from filtered metal vapor plasma. The substrate bias during the metal ion etching stage is typically setup at 1000 volts resulting in acceleration of metal ions across the plasma sheath to kinetic energies ranging from 1 keV for 1^+ metal ions to 4 keV for 4^+ metal ions. The high voltage metal etching treatment results in formation of ultra-thin surface layer enriched with the coating-forming atoms which dramatically increases the adhesion strength at the coating-to-substrate interface[1,2,4,16,29,57,58]. Note that unfiltered DCAS does not provide sufficient improvement of the adhesion toughness because macroparticles deposited by direct cathodic arc process do not have good adhesive bonding to the substrate[4,11,12,58,27]. In a conventional magnetron sputtering process this stage cannot be provided because of very low ion-to metal atoms arrival ratio in a magnetron sputtering process. The ion-to-atom arrival ratio in HIPIMS process may exceed 30% which makes this process capable of providing high voltage metal ion etching stage; however, the LAFAD process still has an advantage of being capable of generating 100% ionized metal vapor plasma which is beneficial for metal ion etching. After ion etching, the bias voltage is reduced and the coating deposition process is performed in a reactive gaseous plasma atmosphere. The substrate bias in LAFAD processing of hard ceramic coatings typically ranging from 40 to 100 volts, in most cases does not exceed 40 volts to reduce the substrate temperature, during coating deposition stage, and intrinsic stresses in the coatings.

The comparison of conventional TiAlN monolithic coating deposited by DCAD process vs. LAFAD TiAlN coating is shown in Figure 5. It can be seen that the LAFAD coating provides virtually no surface defects while coating deposited by the DCAD process has a large macroparticles density, microdelaminations and pin-holes, which is especially important in such applications as corrosion resistant coatings, diffusion barrier layers and high temperature oxidation resistant coatings[1,2,4,14,24,25,35,41,59,60].

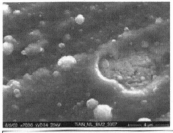

CONVENTIONAL

TiAlN
(3.5 um)

LAFAD

Figure 5. TiAlN monolithic ceramic coating deposited by conventional DCAD process vs. LAFAD coating.

Vacuum cathodic arc evaporation process does not require additional gaseous atmosphere in a vacuum chamber for sustainable operation of vacuum cathodic arc sources[1,2,4,27,58]. The LAFAD process of deposition the hydrogen-free carbon diamond-like carbon coatings (DLC) is performed without adding any plasma carrier gases such as argon[1,2,16]. The continuous carbon vacuum arc is operating at vacuum better than 10^{-5} torr in a self-generated carbon vapor as a plasma carrier gas. The thermal management of substrates during deposition of DLCs has to be able to remove large thermal flux conveyed to the substrates from the carbon plasma stream[1,16]. The energy of carbon ions bombarding the surface of the growing coating cannot exceed 150 eV. Both overheating and ion bombardment by high energy ions can destroy the diamond-like structure and graphitize the coating[1,4,15,16-20]. The HRTEM of 6 μm thick DLC deposited on water-cooled aluminum disk substrate by the LAFAD process[16] is shown in Figure 6. A relatively thick intermediate zone consisting of a mixture of aluminum-carbon atoms can be seen at the carbon DLC-aluminum substrate interface. The formation of this zone is a result of mixing carbon and aluminum atoms under conditions of condensation with intense ion bombardment[16,57]. The hardness of this coating exceeds 70 GPa as illustrated by hardness measurement curves in Figure 7. This hardness is comparable to the hardness of synthetic diamonds. The filtered arc DLCs also have exceptional corrosion resistance owed by their dense and pin-hole free morphology and chemical inertness[1,67]. The surface displacement-loading curve in nano-hardness indentation has demonstrated almost full recovery which shows high elasticity of hydrogen-free DLCs deposited by the LAFAD process[16]. Different morphology and structure of DLC coatings, deposited by the LAFAD, can be achieved by varying the processing parameters such as substrate bias and currents of the primary DCAD sources. Characterization of LAFAD DLCs of different morphologies resulted in a wide variety of functional properties vs. LAFAD process parameters are discussed in details in[1,15-20]. The

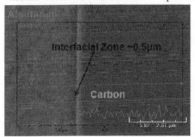

Figure 6. HRTEM cross-section of 6 mm thick DLC deposited by LAFAD process on water-cooled aluminum disk substrate (courtesy of A.Voevodin).

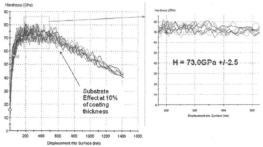

Figure 7. Nanohardness measurement of DLC film shown in Figure 6.

$H = 73.0 GPa +/-2.5$

many applications of LAFAD DLC coatings include medical and dental instruments, cutting and forming tools, machine parts, magnetic media and optical coatings among others.

LAFAD process is capable of depositing coatings with multi-elemental compositions and with complex coating architectures. The multilayer metal-ceramic coatings can be deposited by periodically changing the gaseous plasma atmosphere. For instance, multilayer Ti/TiN coating with Ti metallic sublayers having a thickness ~50 nm followed by TiN ceramic sublayers having a thickness ~500 nm is shown in Figure 8. This coating is deposited by LAFAD process when nitrogen is added to argon every 20 minutes for deposition of TiN sublayer while during 5 minutes between depositions of nitride sublayers the metallic titanium sublayers are deposited in pure argon gas atmosphere[23,24,36]. When targets made of different metals are installed in the primary cathodic arc sources of LAFAD plasma source the nanolayered

Thickness of TiN sublayers

First film	:	.4463 µm
Second film	:	.3988 µm
Third film	:	.5041 µm
Fourth film	:	.5761 µm
Fifth film	:	.5351 µm
Sixth film	:	.4641 µm
Seventh film	:	.4962 µm
Eight film	:	.5282 µm
Ninth film	:	.4884 µm
Tenth film	:	.6646 µm
Eleventh film	:	.7046 µm
Total Thickness :		**5.806 µm**

Figure 8. CALO-test wear-scar image of multilayer Ti/TiN coating deposited by LAFAD process.

a   b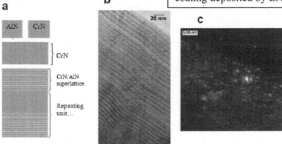
c

Figure 9. Nanolaminated CrN/CrAlN coating deposited by LAFAD source with primary DCAD sources equipped with Al and Cr targets: a-coating design; b- HRTEM of the coating cross-section; c- ~1 nm size crystals incorporated in nanolaminated coating sublayers.

ceramic coating can be deposited by periodically exposing the rotating substrates to the metal plasma generating by two opposite DCAD sources. Figure 9 shows CrAlN nano-structured coating complex architecture as deposited by LAFAD process[35,40]. This coating consists of CrN sublayers followed by CrN/AlN superlattice segments. CrN sublayers were deposited when only one primary DCAS, with a chrome target, was working while the opposite DCAS, with aluminum target, was shut off. The CrN/AlN nano-multilayer segments were deposited when rotating substrates were exposed, in turn, to both Cr and Al vapor plasma flows generated by a LAFAD unidirectional source with opposite primary DCAS equipped with chromium and aluminum targets. Thickness of each individual sublayers in CrN-AlN nano-multilayer segments ranged from 3 to 5 nm while nitride nano-crystals in the CrN-AlN nano-multilayer segments did not exceed 1 nm.

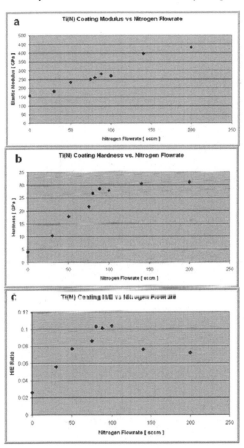

Figure 10. Mechanical properties of sub-stoichiometric TiN coating deposited by LAFAD process vs. nitrogen flowrate.

Ultra-fine grains and almost defectless morphology of LAFAD ceramic coatings are in many cases result in an increase of hardness and elastic modulus while at the same time having increased internal stress and decreased toughness typical for hard ceramic coatings[33,52]. Both chemical composition and coating architecture have been used to optimize the mechanical properties of LAFAD hard coatings. The increase in the Ti layer thickness results in a decrease in the internal stress in both the TiN and Ti layers, and an increase in the grain size and crystallinity of both TiN and Ti phases[52]. Adding Si to TiN coating by using a composite Ti-Si targets for primary DCAD sources of the LAFAD plasma source allows deposition of multiphase coatings consisting of TiN, Si_3N_4 and TiS_2 phases and control of intrinsic stress, grain size and other mechanical properties of the coatings[53,54]. It was found that by adding Si into TiN coatings reduces the grain size significantly from 16.9 to 5.8 nm, changes the orientation from (111) to (220) preferred orientation, and increases the hardness and Young's modulus from 33 and 376 GPa to 51 and 449 GPa, respectively. XPS and XRD results show that the Si/Ti atomic ratio in the coatings is 0.17 and the deposited TiSiN coatings consist of nanosized TiN grains encapsulated by amorphous Si_3N_4 layer, corresponding to the superhardness of the TiSiN coatings. The high plasma density, ion energy, and ion reactivity of the filtered cathodic arc plasma contribute to the formation of the nanocomposite TiSiN coatings at low temperature and low N_2 partial pressure as it does in PEMS process[9,53,54,56].

TiN coating properties can be also controlled by varying nitrogen content[55]. The mechanical properties of sub-stoichiometric TiN coatings deposited by LAFAD process with different nitrogen flowrate are shown in Figure 10. It can be seen that hardness and elastic modulus increased with the increase of nitrogen flowrate reaching the maximum in stoichiometric coating composition. In a contrast, the H/E ratio associated with coating toughness reached the maximum in under-stoichiometric coating composition[51].

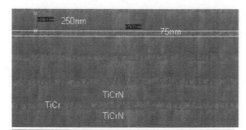

Figure 11. Multilayer TiCr/TiCrN wear and corrosion resistant coating deposited by LAFAD process.

Multilayer TiCrN/TiCr cermet coatings deposited by the LAFAD process have demonstrated a superior corrosion resistance along with low stresses, exceptionally good adhesion and high toughness[1,12,28,44,47]. The cross-section of this coating, having Ti50Cr12N38 composition, is shown in Figure 11[49]. The results of salt fog simulated corrosion testing of M50 steel with 6 µm LAFAD TiCrN/Ti coatings have demonstrated almost a complete elimination of pitting corrosion, as can be seen in Figure 12[49,28].

Further improvement was achieved by using a 2-segment TiCrN/Ti+TiBCN coating deposited by a hybrid LAFAD+UBM process[44,47]. The hybrid FAPSID surface engineering system configuration is capable of depositing LAFAD+UBM coatings with modulated concentrations of one or more components by using LAFAD source in a magnetic shuttering mode. In this mode the deflecting magnetic field is activated and de-activated with repetition frequency ranging from 0.1 to 10 Hz and

Figure 12. Carburized Pyrowear 675 discs after salt fog testing: left-uncoated; right- with TiCr/TiCrN coating shown in Figure 11.

with a controlled duty cycle. When the deflecting magnetic field is ON, the TiBC coating is deposited by a hybrid LAFAD+UBM process. When deflecting magnetic field is OFF only B$_4$C sublayer is depositing by UBM source which operates continuously. The HRTEM cross-sectional image of modulated TiBC coating is shown in Figure 13[47]. It can be seen that the bi-layer period in this coating does not exceed 4 nm. The hardness and elastic modulus of TiBC coating deposited by the hybrid LAFAD+UBM process in a magnetic shuttering mode depends on the modulating period of Ti-enriched sublayer and Ti concentration in a coating achieving superhard properties with hardness of about 45 GPa at optimized titanium content[47].

Figure 13. HRTEM cross-sectional image of the TiBC top segment nanolaminated architecture: a- complete top segment architecture; b-magnification of the top end of the TiBC segment.

The same type of the coating also demonstrated extremely high wear and corrosion resistance in hot perfluoropolyalkylether (PFPAE) corrosion under high load rolling conditions establishing dramatic differences between TiBC coating configurations performance and uncoated Pyrowear 675 samples[28,44,47]. Initial tests were conducted for 2 h at standard test conditions with repeated results, as shown in Figure 14. Testing durations of 2 h demonstrated relatively large, 4–8 μm deep, plastic wear track deformations occurring in the uncoated sample in comparison to the coated samples, which showed only polishing of coating in the form of a local reduction in the coating surface roughness. The multi-phase nanostructured TiBC(N) layer deposited by hybrid LAFAD+UBM process have also demonstrated improved wear resistance in high load contact sliding condition which was attributed to combination of low friction and high adhesion toughness of these coatings[28,44,47].

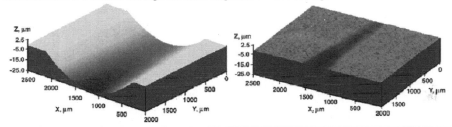

Figure 14. Scanning profilometry image of lubricated sliding wear track, using a 2.0 μm TiCrN/TiCrCN coated ring on an uncoated carburized Pyrowear 675 coupon (left) and on a coupon with TiCrN/TiCrCN+TiBC coating deposited by LAFAD+UBM hybrid process.

The hybrid LAFAD+UBM surface engineering system was used for deposition of both nano-composite and two-segment coatings for cutting tools. The two-segment hard-soft coatings deposited by the LAFAD+MS process, consisted of the first segment hard ceramic coating deposited by the LAFAD source followed by soft, low friction segment deposited by magnetron sputtering deposition process[1,4,58,65,66]. The hard bottom segment was made of TiN, TiAlN, TiC, CrN or TiCN. These ceramic coatings deposited by LAFAD process both multilayers similar to that shown in Figure 8 and monolithic ceramic layers have hardness ranging from 20 to 40 GPa. Various materials were deposited by magnetron sputtering as a top soft low friction coating segment: WS$_2$, MoS$_2$ and soft carbon coating[65]. The WS$_2$ coatings were also deposited by cathodic arc evaporation of tungsten in Ar+H$_2$S reactive gas atmosphere. The results of testing different cutting tools with 2-segment hard-soft coating have demonstrated increased wear resistance and durability of these tools compared to cutting tools having only hard wear resistant coating[65].

Further improvement of the hard-soft coatings for cutting tools was achieved by using nanocomposite coating systems, employing a soft phase incorporated into the hard phase during LAFAD and a hybrid LAFAD+MS deposition processes[65]. The following nanocomposite mixed hard-soft coatings were deposited and tested for their performance in cutting tools applications: TiC+C, TiN+Ag and TiN+MoS2. The TiN+MoS$_2$ coatings were deposited both by co-deposition of TiN and MoS$_2$ during simultaneous operation of LAFAD source with titanium primary arc targets and MoS2 by magnetron sputtering respectfully. It was also deposited by operating of cathodic arc sources with Ti and MoS$_2$ targets. TiC+C coatings were deposited by LAFAD process using Ar+CH$_4$ as a reactive gas atmosphere. The mixed TiN+Ag hard-soft coatings were deposited by cathodic arc evaporation of segmented Ti+Ag primary cathodic arc targets of the LAFAD plasma source. In addition, a nanocomposite MoN+Cu coating was deposited by co-evaporation of Mo and Cu primary cathodic arc targets installed in opposite sides of the LAFAD plasma source in nitrogen reactive gas atmosphere.

This coating becomes lubricous when undergoes oxidation at high temperatures during cutting[65]. Again, considerable improvement in the tool life was demonstrated.

The hybrid LAFAD assisted processes were also used for development of the carbon and boron contained coatings, which were deposited by LAFAD+MS process for cutting tools applications[66]. In these processes titanium and carbon cathodes were used to generate arc plasma. Boron and titanium diboride cathodes were used in the magnetron sources to produce boron containing coatings. The substrates were mounted on a variable speed rotary (two-axis) substrate holder that can be biased to a desired voltage using either a DC or RF power. These coatings have demonstrated improved hardness ranging from 34 GPa to superhard coatings of 65 GPa. The superhard TiBN single layer coating have demonstrated outstanding performance in high speed end milling of hardened tool steels (H-13). It has demonstrated a tool life increase of exceeding 800% over uncoated WC/Co and 25% higher than the second best candidate: multilayer TiBN/TiN[66].

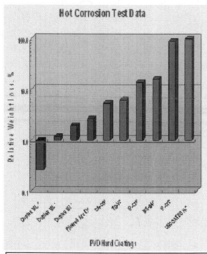

Figure 15. Performance of TiN, TIBN, and TICN multilayer coatings deposited by the filtered-arc method, in a simulated pressure die-casting erosion_corrosion environment, as compared to other commercial coatings: duplex ML1- ionitriding+TiBCN/TiN multilayer (LAFAD); duplex ML2- ion-nitriding+TiN/TiCN multilayer (LAFAD); duplex ML3- ionitriding+Ti/TiN multilayer (LAFAD); CA-CrN, direct cathodic arc deposition; VC- Toyota-diffusion process; IP-CrN, ion-plated CrN; MS-B4C, magnetron-sputtered B_4C; and IP-CrC, ion-plated CrC.

Multi-component LAFAD coatings have demonstrated superior high temperature corrosion and oxidation resistance in a chemically aggressive environment, which makes LAFAD and hybrid LAFAD+UBM and LAFAD+EBPVD processes technologies of choice for wide variety of industrial applications. Duplex coatings consisting of ionitrided layer followed by Ti/TiN/TiNC/TiBNC multilayer, multi-elemental, coatings have demonstrated superior high temperature corrosion resistance in a contact with molten Al-16% Si alloy environment simulating aluminum foundry process using steel made die casting dies[23-25]. The results of weight loss experiments from the erosion-corrosion test in molten aluminum alloy, shown in Figure 15, demonstrate superior performance of LAFAD coating vs. other coatings and surface treatments. The plot is normalized with respect to the weight loss of an uncoated H-13 pin, by assigning this a value of 100, and the results are plotted on a logarithmic scale. Lower weight loss indicates a better performance. The multilayer coatings show a weight loss which is nearly two orders of magnitude smaller than that of the uncoated H-13 pin. It was also demonstrated that multilayer coatings deposited by LAFAD process show minimum ejection forces and thus the minimum adhesion tendency[23-25].

Another example of using LAFAD ceramic coatings for high temperature oxidation resistance applications are coatings for the SOFC interconnects (SOFC IC) made of inexpensive ferritic stainless steels[42,45]. Near defectless morphology and dense structure of LAFAD coatings resulted in their ability to inhibit high temperature oxidation by eliminating pin holes, voids and porosity as well as reducing the inward/outward mobility of cations/anions. The multielemental oxi-ceramic coatings of

CoMnTiCrAlYO system deposited on ferritic steel plates by the LAFAD process have demonstrated its ability to slow the growth of thermally grown oxide scale (TGO) while securing high electrical conductivity and almost completely blocking the chrome outward diffusion, which is poisoning the SOFC cathode[41]. Two-segment coatings were deposited by LAFAD and hybrid LAFAD+EBPVD processes. The bottom segment consists of TiCrAlO layer deposited by LAFAD source with primary DCAD sources equipped with multi-elemental TiCrAlY targets. The elemental composition of TiCrAlYO coating deposited in this process is replicating the cathode target elemental composition with high accuracy[48]. The LAFAD oxi-ceramic coatings, having excellent diffusion barrier properties, were used as the bottom segment of a two-segment coating architecture with a CoMnO top segment deposited by a hybrid LAFAD+EBPVD coating deposition process in the FAPSID surface engineering system[48]. The diffusion barrier performance of this dual-segment coating is illustrated in Figure 16. It can be seen that even a 0.3 μm bottom TiCrAlO adhesive and diffusion barrier segment provides excellent oxidation protective properties, inhibiting the TGO growth after 1500 hrs of oxidation. In a

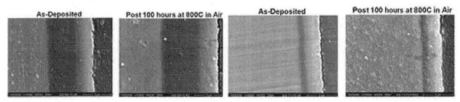

Figure 16. SEM/EDS cross sections of dual segment coating before and after oxidation in 800C air. Top left: 3.0 um CrAlYO + 1 um MnCoO: no chromia TGO scale growth showing no substantial change in coating structure after high temperature exposure. Top right: 0.3 um CrAlYO + 1 um MnCoO: still no chromia TGO scale growth and no substantial change in coating structure after high temperature exposure.

further development the nanocomposite coating of (Co,Mn)TiCrAlYO system were deposited on ferritic steel substrates by the LAFAD process using primary DCAS targets made of composite (Co,Mn)TiCrAlY alloy[42]. Excellent diffusion barrier properties of LAFAD coatings were also used in deposition of alumina interlayer between bondcoat and top ceramic coating of the TBC system for turbine blade applications. The LAFAD defectless alumina layer having very low oxygen diffusivity was used to slow the formation of TGO scale on bondcoat–to–topcoat interface and demonstrated improvement in stability of TBC against buckling in thermal cycling testing[59,60].

The LAFAD multilayer metal-ceramic coatings similar to that shown in Figure 8 are currently applied for the mass production of dental instruments such as handle and ultra-sonic scalers and curettes[36-39]. Thickness of these coatings deposited on high chromium stainless steel ranged from 2 to 5 microns. This LAFAD process allows to retain the sharpness of the instruments as the radius of curvature at the cutting edge does not exceed coating thickness, which is critically important for this application. The intrinsic stresses and hardness of these coatings is controlled via the ratio of titanium metallic sublayer thickness to ceramic TiN sublayer thickness in these multilayer coatings[36, 52]. The hardness of this coating ranges from 25 to 35 GPa, which allows multiple usage of scalers in periodontal dental operations without re-sharpening. On the other hand, uncoated instruments require re-sharpening after its use for each patient. The absence of macroparticle inclusions and ultra-fine coating morphology are also adding exceptional abrasion resistance which was demonstrated by subjecting dental instruments to vibratory treatment in a contact abrasion media[37-39].

ULTRA-THICK COATINGS AND EFFICIENCY OF LAFAD PROCESS

A record high metal vapor plasma transport efficiency of unidirectional LAFAD source allows deposition of ultra-thick ceramic and cermet coatings of different architectures and compositions as erosion and abrasion wear protection and corrosion resistance coatings for components operating in extreme conditions such as aircraft parts (compressor blades and helicopter rotorblades) and mining tools[49-51]. The thickness of TiN based coatings deposited by one unidirectional LAFAD source on substrates installed in 0.5 m diameter rotary table with single rotation (one side coating) ranged from 20 to 60 μm with deposition rate exceeding 5 μm/hr. In addition, two-segment coatings utilizing either monolithic I,J coatings or multilayer G coating as a bottom segment interfacing the substrate and nano-multilayer K_1 top coating segment were also produced. The cross-section of a two-segment GK_1 coating,

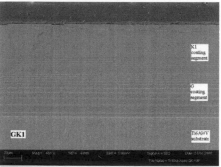

Figure 17. Two-segments nano-microlaminated ultra-thick Ti/TiN coating deposited by LAFAD process.

deposited during two consecutive coating runs is shown in Figure 17[50,51]. The bottom microlaminated 10-layer G-coating bottom segment was deposited during 10 hrs deposition run with 15 min deposition of metallic titanium sublayers in argon, followed by 45 min deposition of TiN sublayers in nitrogen reactive gas atmosphere. The top nano-microlaminated 40-layer K_1-coating segment was deposited during 11 hrs run with 2 min deposition of metallic titanium sublayers in argon followed by 13 min deposition of TiN sublayers in nitrogen as a reactive gas atmosphere. Good mechanical stability and low stresses in ultra-thick TiN base coatings may be explained by the role of thickness-dependent gradients of point defect density[32].

The current of the primary titanium cathode targets during deposition of G and K1 coatings was 200 amperes resulting in large melting areas surrounding the cathodic arc spots. The maximum deposition rates of LAFAD process has been achieved when the target surface reaches melting temperature in the large area surrounding cathodic arc spots. This condition is achieved during deposition of G and K coatings as can be seen from the Figure 18 photograph depicting the surface of a titanium cathode target exposed to 200 amperes arc current during the deposition of G coating as seen from the titanium target surface. The deposition rates of G and K_1 coatings deposited by LAFAD process with 200 amperes of the primary DCAS current vs. process time in comparison with deposition rate of TiN based coating deposited at lower cathode current of 140 amperes are shown in Figure 19. It can be seen that the deposition rate of TiN coating reaches 10 μm/hr at the beginning of LAFAD deposition process at 200 amperes of primary DCAS sources when the target surface temperature reaches its maximum value near the melting temperature of titanium in areas close to the cathodic arc spots. This deposition rate is almost three times greater than deposition rate of

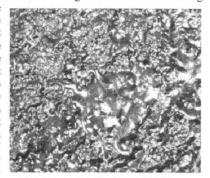

Figure 18. Surface of titanium cathodic arc target subjected to arc evaporation process with 200 amperes arc current.

the TiN coating deposited by a LAFAD source with 140 amperes of the primary DCAS currents. The deposition rate is reduced when the water cooled targets get thinner, resulting in improved water cooling efficiency and reduced integral temperature of the target surface. The uniformity of the coating thickness distribution during deposition of ultra-thick TiN based coatings, with vertical magnetic rastering, is about +/-15% over 12 inches tall coating area and without magnetic rastering, over 6 inches tall deposition area[22,29-31,49]. Using this data the ion current transmission rate in LAFAD process can be calculated. Assuming the value of the average titanium ion charge as $<z_{Ti}>=2.2$[4], the maximum coating deposition rate of 10 µm/hr, and a coating area of 500 cm diameter × 15 cm tall as in G- and

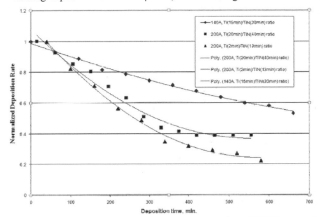

K-coating processes, the ion current transmitted from two primary DCAD sources of LAFAD plasma source can be estimated at 16 amperes which corresponds to 4% of the total ion current generated by two DCAS with 200 amperes arc current each. Taking into account that maximum ion current generating in vacuum cathodic arc process is less than 8% of the total arc current[4] it can be seen that LAFAD process is capable of transporting more than a half of the total ion current generating by primary vacuum arc sources toward substrates to be coated along the curvilinear deflecting magnetic field. In contrast, the conventional filtered arc deposition processes have efficiency of ion current transmission less than 1%[4,11,12,27,58]. Plasma transport

Figure 19. Normalized deposition rates of Ti/TiN multilayer coatings with different architectures deposited by unidirectional dual arc LAFAD metal vapor plasma source on rotating (one side coated) substrates.
Maximum Deposition Rates: 140A, Ti(15min)/TiN(30min)- 2.7um/hr; 200A, Ti(20min)/TiN(40min)- 10.7 um/hr; 200A, Ti(2min)/TiN(13min)- 8.2 um/hr.

efficiency of the LAFAD process is at least equal or, for many types of commercially available DCAD sources, exceeds the ion transport efficiency of the unfiltered direct arc deposition process. High efficiency of vapor plasma transport in LAFAD process can be explained by concave topology of the deflecting magnetic field of unidirectional dual arc LAFAD plasma source. It is known that diffusion losses are minimal when plasma is confined in concave magnetic field, while plasma diffusion across magnetic force lines is maximal when plasma is confined in convex magnetic field[3]. On the other hand, much greater diffusion losses of vapor plasma were found in a straight vacuum arc column configuration where the arc was confined in an axi-symmetrical longitudinal magnetic field[2,46].

Ultra-thick ceramic and cermet coatings deposited by the LAFAD process possess relatively high hardness ranging from 18 GPa for laminated coatings to 35 GPa for monolithic coatings. The high hardness of LAFAD coatings was combined with high fracture resistance, making these coatings capable of erosion protection of turbomachinery components such as compressor blades of turbine engines and protectors for helicopter rotorblades, as well as in abrasive wear applications such as mining and road rehabilitation tools. The laminated architecture is a well known approach to increase toughness of ceramics, cermets and composites[34]. It was found that crack trajectories in ultra-thick laminated Me/MeN LAFAD coatings that have a relatively thick metallic interlayers are predominantly lateral, arrested in ceramic sublayers, while restricted by two neighboring metallic sublayers; however, in monolithic ceramic coatings the crack pattern is random[51]. Even better fracture resistance can be obtained by using sub-stoichiometric coatings, as illustrated in Figure 20, showing the cross-section of Rockwell 145 kg load indentation of L coating made of TiAl/TiAlN multilayer vs. sub-stoichiometric TiN coating[51]. It can be seen that density of cracks developed in a central area of indentation subjected to high compressive stress in sub-stoichiometric coating having hardness of about 25 GPa is much

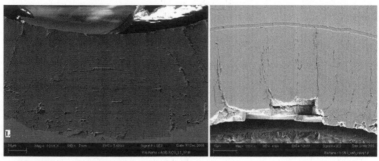

Figure 20. Rockwell (145 kg) indentation cross-sections of ultra-thick TiAl/TiAlN multilayer coating (left) and sub-stoichiometric TiN coating (right) deposited by LAFAD process.

smaller than that of laminated stoichiometric L coating (TiAl/TiAlN multilayer) having about the same hardness and thickness. The combination of high hardness and good fracture resistance in LAFAD ultra-thick ceramic coatings resulted in high erosion resistance. The LAFAD TiN and TiAlN-based ceramic and cermet coatings with thicknesses up to 100 µm, can still provide a smooth surface even with a roughness of RMS<1µm. Thick TiN based coatings deposited by the LAFAD process, both microlaminated Ti/TiN cermet and monolithic TiN ceramics, can provide an order of magnitude improvement of erosion resistance under conditions of impacts with runway sand particulate flow at the speed up to 1200 fps[50]. These properties of the LAFAD process make it an attractive alternative to replace conventional plasma PVD processes for a wide range of applications in turbomachinery.

CONCLUSIONS

The unidirectional dual arc LAFAD vapor plasma source has been characterized as a generator of high-density ion current and mass flow with 100% ionized metal vapor and more than 50% ionized gaseous plasma component. The productivity of the LAFAD plasma source is comparable to, or exceeds, the productivity of conventional DCAD sources and magnetron sputtering sources. For example, one LAFAD source can deposit TiN-based coatings on substrates installed on the 0.5 meter diameter rotating turntable of the industrial size batch coating chamber with an average productivity exceeding 5 µm/hr, yet provide a uniform coating thickness distribution a ~+/-15% over the 12 inches tall deposition area. The efficiency of the LAFAD process is defined by its ability to transport more than half of the total ion current generated by vacuum arc process from the primary DCAS evaporating targets toward substrates to be coated along curvilinear deflecting magnetic field. The LAFAD vapor plasma transport efficiency exceeds more than four times the transport efficiency of the conventional FCAD process and, in many cases, even exceeds the direct cathodic arc deposition process. Cermet and ceramic coatings of different multielemental multiphase compositions and architectures are featuring near defectless morphology, smooth surface and extremely high adhesion. The operating range of the LAFAD process allows combining it with other PVD processes such as magnetron sputtering and EBPVD, as well as low pressure plasma assistant CVD process, forming hybrid LAFAD plasma assisted surface engineering technologies. Both micro- and nano-laminated ceramic and cermet coatings and nanocomposite coatings deposited by LAFAD process have demonstrated superior functional properties in various applications including wear resistant coatings for cutting and forming tools and medical instruments, erosion resistant coatings for turbomachinery components, high temperature oxidation resistant coatings for SOFC, tribological coatings for gears and bearings operating in extreme conditions. These properties of the LAFAD process make it an attractive alternative to replace conventional plasma PVD processes for a wide range of applications.

REFERENCES

[1] V.I. Gorokhovsky, et al., Report on the Investigation of Physics and Chemistry of Chemically Active Plasma Flow for Development of New Synthesis Methods for Composite Superhard Materials and Deposition of Coatings, Report No. 0287-0034250, Institute for Superhard Materials of the Ukrainian Academy of Sciences, Kiev, Ukraine (1983).

[2] V.I. Gorokhovsky, V.P. Polistchook and I.M. Yartsev, Process in Plasma-Arc Installation for Vacuum Depositions, Part I: Plasma Generation; Part II: Plasma Propagation, *Surf Coat Tech*, **61**, 101 (1993).

[3] N. A. Krall and A. W. Trivelpiece, Principles of Plasma Physics, McGraw-Hill Book Company, New York, USA (1973).

[4] R.L. Boxman, D.M. Sanders, and P.J. Martin, *Handbook of Vacuum Arc Science and Technology*. Noyes Publications, Park Ridge, New Jersey, USA (1995).

[5] J. Musil, Hard Nanocomposite Films Prepared by Reactive Magnetron Sputtering, in *Nanostructured Coatings*, edited by A. Cavaliero and J. De Hosson, Springer Publishing Co. (2006).

[6] S.M. Rossnagel, Directional and Preferential Sputtering-Based Physical Vapor Deposition, *Thin Solid Films* **263**, 1-12 (1995).

[7] U. Helmersson, M. Lattermann, J. Bohlmark, A.P. Ehiasarian, and J.T. Gudmundsson, Ionized Physical Vapor Deposition (IPVD): A Review of Technology and Applications, *Thin Solid Films*, **513** 1-24 (2006).

[8] M. Klingenberg, J. Arps, R. Wei, J. Demaree, and J. Hirvonen, Practical Applications of Ion Beam and Plasma Processing for Improving Corrosion and Wear Protection, *Surf Coat Tech*, **158-159**, 164-169 (2001).

[9] R. Wei, E. Langa, C. Rincon, and J.H. Arps, Deposition of Thick Nitrides and Carbonitrides for Sand Erosion Protection, *Surf Coat Tech*, **201**, 4453-4459 (2006).

[10] Q. Luo, W.M. Rainforth, and W.-D. Münz, Wear Mechanisms of Monolithic and Multicomponent Nitride Coatings Grown by Combined Arc Etching and Unbalanced Magnetron Sputtering, *Surf Coat Tech*, **146-147**, 430-435 (2001).

[11] A. Aksenov, V. A. Belous, V. G. Padalka, and V. M. Khoroshikh, *Soviet Journal of Plasma Physics*, **4**, 425 (1978).

[12] D. Baldwin and S. Falabella, Technical Proceedings of the Annual Meeting of Society of Vacuum Coaters (1995).

[13] V.I. Gorokhovsky, T. Kurtinina, and D. Otorbaev, in *Collection of Scientific Papers on Wear Resistant and Protective Coatings*, edited by M.A. Voronkin, Institute of Superhard Materials, Academy of Sciences of the UkrSSR, Kiev [in Russian], (2001).

[14] K. Gordashnik, V. Gorokhovsky, and B. Uryukov, Study of Corrosion Stability of Medical Instruments with TiN-Coatings, Proceedings of the International Ion Engineering Congress, ISIAT 83, Institute of Electrical Engineers, Kyoto, Japan (1983).

15. Z. Has, S. Mitura, and V.I. Gorokhovsky, The System for Depositing Hard Diamond-Like Films onto Complex-Shaped Machine Elements in an R.F. Arc Plasma, *Surf Coat Tech*, 47, 106-112 (1991).

16. N. Novikov, V.I. Gorokhovsky, and B. Uryukov, Superhard i-C Coatings Used in Complex Processes of Surface Strengthening of Tools and Machine Parts, *Surf Coat Tech*, **47**, 770 (1991).

17. V. Aleshin, V.I. Gorokhovsky, V. Malnev, E. Pugach, and A. Smekhnov, Structure and Properties of iC-Coatings Deposited from Plasma Phase, *Archive Nauki o Materialach* (Poland), 7, 105-110 (1986).

18. V. Aleshin, V.I. Gorokhovsky, V. Strel'nitsky, and B. Uryukov, Interaction at the Interface Between Carbon Films and Mmetals, *Soviet Journal of Superhard Materials*, 9, 19-23 (1987).

19. V.I. Gorokhovsky, Y. Leshchiner, and A. Malyshev, Wear Characteristics of HSS with TiN, CrN and i-C Coating Deposited from Plasma Phase, Proceedings of the International Seminar on Vacuum Coatings for Wear Surfaces of Tools and Machine Parts Using Plasma, Warsaw, Poland (1987).

20. A. Lysenko, V.I. Gorokhovsky, Y. Nikitin, and V. Poltoratsky, Molecular Model of Diamond-Like Carbon Structure Synthesized under Low Pressure, Transactions of the Academy of Sciences of the UkrSSR [in Russian] (1987).

21. V.I. Gorokhovsky, Vacuum Cermet Coatings on Coiled Materials, Proceedings of the Fourth International Conference on Vacuum Web Coating, edited by R. Bakish, Reno, Nevada, USA (1990).

22. V.I. Gorokhovsky, R. Bhattacharya, and D.G. Bhat, Characterization of Large Area Filtered Arc Deposition Technology: Part I-Plasma Processing Parameters, *Surf Coat Tech*, **140**, 82-92 (2001).

23. V.I. Gorokhovsky, D.G. Bhat, R. Bhattacharya, A.K. Rai, K. Kulkarni, and R. Shivpuri, Characterization of Large Area Filtered Arc Deposition Technology: Part II-Coating Properties and Applications, *Surf Coat Tech*, **140**, 215-224 (2001).

24. D.G. Bhat, V.I. Gorokhovsky, R. Bhattacharya, R. Shivpuri, and K. Kulkarni, Development of a Coating for Wear and Cracking Prevention in Die-Casting Dies by the Filtered Cathodic Arc Process, Transactions of the North American Die Casting Association, Twentieth International Die Casting Congress and Exposition, Cleveland, Ohio USA (1999).

25. V. Joshi, K. Kulkarni, R. Shivpuri, R.S. Bhattacharya, S.J. Dikshit, and D. Bhat, Dissolution and Soldering Behavior of Nitrided Hot Working Steel with Multilayer LAFAD PVD Coatings, *Surf Coat Tech*, **146 –147**, 338–343 (2001).

26. P.J. Martin and A. Bendavid, Review of the Filtered Vacuum Arc Process and Materials Deposition *Thin Solid Films*, **394**, 1-15 (2001).

27. D.M. Sanders and A. Anders, Review of Cathodic Arc Deposition Technology at the Start of the New Millennium, *Surf Coat Tech*, **133-134**, 78-90 (2000).

28. www.arcomac.com.

29. V.I. Gorokhovsky, US Patent No. 7,300,559 (November 27, 2007).

30. V.I. Gorokhovsky, C. Bowman, D. VanVorous, and J. Wallace, Deposition of Various Metal, Ceramic and Cermet Coatings by an Industrial-Scale LAFAD Process," *JVST A*, **27**, 1080-1095 (2009).

31. V.I. Gorokhovsky, Characterization of Thick Ceramic and Cermet Coatings Deposited by an Industrial-Scale LAFAD Process, *Surf Coat Tech*, **204**, 1216-1221 (2010).

32. H. Kostenbauer, G.A. Fontalvo, M. Kapp, J. Keckes, and C. Mitterer, Annealing of Intrinsic Stresses in Sputtered TiN Films: The Role of Thickness-Dependent Gradients of Point Defect Density, *Surf Coat Tech* **201**, 4777-4780 (2007).

33. K.J. Ma, A. Bloyce, and T. Bell, Examination of Mechanical Properties and Failure Mechanisms of TiN and Ti-TiN Multilayer Coatings, *Surf Coat Tech*, **76-77**, (1995) 297-302.

34. W. Lee and W.J. Clegg, "The Deflection of Cracks at Interfaces, *Key Engineering Materials*, **116-117**, 193-208 (1996).

35. R.J. Smith, C. Tripp, A. Knospe, C.V. Ramana, A. Kayani, V.I. Gorokhovsky, V. Shutthanandan, and D.S. Gelles, Using CrAlN Multilayer Coatings to Improve Oxidation Resistance of Steel Interconnects for Solid Oxide Fuel Cell Stacks, *J Mater Eng Perform*, **13**, 295-302 (2004).

36. V.I. Gorokhovsky, B. Heckerman, P. Watson, and N. Bekesch., The Effect of Multilayer Filtered Arc Coatings on Mechanical Properties, Corrosion Resistance, and Performance of Periodontal Dental Instruments, *Surf Coat Tech*, **200**, 5614-5630, (2006).

37. V.I. Gorokhovsky, B. Heckerman, and T. Brown, Advanced Surface Engineering Technology for Endodontic Instruments and Related Applications, *Roots*, **Number 2**, 40-50 (2006).

38. V.I. Gorokhovsky, B. Heckerman, and T. Brown, Advanced Surface Engineering Technology for Endodontic Instruments and Related Applications, *International Dental Tribune*, Endo Supplement, (2006).

39. www.am-eagle.com

40. P.E. Gannon, C.T. Tripp, A.K. Knospe, C.V.R. Ramana, M. Deibert, R.J. Smith, V.I. Gorokhovsky, V. Shutthanandan, and D. Gelles, High-Temperature Oxidation Resistance and Surface Electrical Conductivity of Stainless Steels with Filtered Arc Cr-Al-N Multilayer and/or Superlattice Coatings, *Surf Coat Tech*, **188-189**, 55-61 (2004).

41. C. Collins, J. Lucas, T.L. Buchanan, M. Kopczyk, A. Kayani, P.E. Gannon, M.C. Deibert, R.J. Smith, D.S. Choi, and V.I. Gorokhovsky, Chromium Volatility of Coated and Uncoated Steel Interconnects for SOFCs, *Surf Coat Tech*, **201**, 4467-4470 (2006).

42. A. Kayani, R.J. Smith, S. Teintze, M. Kopczyk, P.E. Gannon, M.C. Deibert, V.I. Gorokhovsky, and V. Shutthanandan, Oxidation Studies of CrAlON Nanolayered Coatings on Steel Plates," *Surf Coat Tech*, **201**, 4460-4466, (2006).

43. V.I. Gorokhovsky, C. Bowman, P. Gannon, D. VanVorous, A.A. Voevodin, A. Rutkowski, C. Muratore, R.J. Smith, A. Kayani, D. Gelles, V. Shutthanandan, and B.G. Trusov, Tribological Performance of Hybrid Filtered Arc-Magnetron Coatings, Part I, *Surf Coat Tech*, **201**, 3732-3747, (2006).

44. V.I. Gorokhovsky, C. Bowman, P. Gannon, D. VanVorous, A. Voevodin, and A. Rutkowski, Tribological Performance of Hybrid Filtered Arc-Magnetron Coatings, Part II," *Surf Coat Tech*, **201**, 6228-6238 (2007).

45. L.R. Pederson, P. Singh, and X. D. Zhon, Application of Vacuum Deposition Methods to Solid Oxide Fuel Cells, *Vacuum*, **80**, 1066-1083 (2006).

46. V.I. Gorokhovsky, V.P. Elovikov, P.L. Lizunov, and S.A. Pantyukhin, An Induction Method of Determining Conduction-Zone Size in a Vacuum Arc, *High Temp*, **26**, 170-175 (1988).

47. V.I. Gorokhovsky, C. Bowman, P.E. Gannon, D. VanVorous, J. Hu, C. Muratore, A.A. Voevodin, and Y.S. Kang, Deposition and Characterization of Hybrid Filtered Arc-Magnetron Multilayer Nanocomposite Cermet Coatings for Advanced Tribological Applications, *Wear* **265**, 741-755 (2008).

48. V.I. Gorokhovsky, P.E. Gannon, M.C. Deibert, R.J. Smith, A. Kayani, M. Kopczyk, D. VanVorous, Z. Yang, J.W. Stevenson, S. Visco, C. Jacobson, H. Kurokawa and S. Sofie, Deposition and Evaluation of Protective PVD Coatings on Ferritic Stainless Steel SOFC Interconnects, *J Electrochem Soc*, **153**, A1886-A1893 (2006).

49. V.I. Gorokhovsky, J. Wallace, C. Bowman, P.E. Gannon, J. O'Keefe, V. Champagne and M. Pepi, Large Area Filtered Arc and Hybrid Coating Deposition Technologies for Erosion and Corrosion Protection of Aircraft Components, Proceedings of 32nd International Conference & Exposition on Advanced Ceramics & Composites, Daytona Beach, Florida USA (2008).

50. V.I. Gorokhovsky, C. Bowman, J. Wallace, Dave VanVorous, John O'Keefe, Victor Champagne, Marc Pepi, Widen Tabakoff, LAFAD Hard Ceramic and Cermet Coatings for Erosion and Corrosion Protection of Turbomachinery Components, Proceedings of ASME Turbo Expo 2009: Power for Land, Sea and Air, Paper No. GT2009-59391, Orlando, Florida USA (2009).

51. V.I. Gorokhovsky, Characterization of Thick Ceramic and Cermet Coatings Deposited by an Industrial-Scale LAFAD Process, Presentation at ICMCT, San Diego, California USA (2008) and published in *Surf Coat Tech*, **204**, 1216-1221 (2010).

52. Y.H. Cheng, T. Browne, B. Heckerman, J.C. Jiang, E I. Meletis, C. Bowman, and V.I. Gorokhovsky, Internal Stresses in TiN/Ti Multilayer Coatings Deposited by Large Area Filtered Arc Deposition, *J Appl Phys* **104**, 093502 (2008).

53. Y.H. Cheng, T. Browne, B. Heckerman, J.C. Jiang, E.I. Meletis, C. Bowman, and V.I. Gorokhovsky, Influence of Si Content on the Structure and Internal Stress of the Nanocomposite TiSiN Coatings Deposited by Large Area Filtered Arc Deposition, *J Phys D: Appl Phys* **42** 125415 (2009).

54. Y.H. Cheng, T. Browne, and B. Heckerman, Nanocomposite TiSiN Coatings Deposited by Large Area Filtered Arc Deposition, *J Vac Sci Tech A*, **27**, 82-88 (2009).

55. A. Sue, Development of Arc Evaporation of Non-Stoichiometric Titanium Nitride Coatings, *Surf Coat Tech*, **62**, 115-120 (1993).

56. V.P. Swaminathan, R. Wei, D. Gandy, Nanotechnology Coatings for Erosion Protection of Turbine Components, ASME Paper No. GT2008-50713 (2008).

57. V.I. Gorokhovsky, A. Zhiglinsky, V. Kuchinsky, and V. Fomichev, Kinetics of Transitional Layer Forming with Precipitation of Surfaces from Ion Beams, *Surface: Physics, Chemistry, Mechanics*, **Number 2** (1991).

58. V.I. Gorokhovsky and D.G. Bhat, Principles and Applications of Vacuum Arc Plasma-Assisted Surface Engineering Technologies, Proceedings of the GOI-UNDP International Workshop on Surface Engineering and Coatings, Bangalore, India, edited by I. Rajagopal, K.S. Rajam, and R.V. Krishnan, Allied Publishers, Ltd., Mumbai, India (1999).

59. R. Bhattacharya and B. Majumbar, Improved Thermal Barrier Coating System Based on a Cathodically Deposited Alpha Alumina Sublayer, Report No. AFRL-ML-WP-TR-2001-4099, UES, Inc., Dayton, Ohio USA (2001).

60. A.K, Rai, Development of Protective Coatings for Single Crystal Turbine Blades, Final Report prepared for The U.S. Department of Energy (Contract No. DE-FG02-03ER83809), UES, Inc. Dayton, Ohio USA (2006).

61. A.K. Rai, J. Nainaparampil, M. Massey, R.S. Bhattacharya, L.J. Gschwender, C.E. Snyder, Jr., Interaction of Fomblin® Lubricant with Surface Nitrided and/or Coated Bearing (M50) Steel, *Lubric Sci*, **21**, 305-320 (2009).

62. S.J. Dixit, A.K. Rai, R. Bhattacharya, and S. Guha, Characterization of AlN Thin Films Deposited by Filtered Cathodic Arc Process, *Thin Solid Films*, **398-399**, 17 (2001).

63. R. Bathe, R.D. Vispute, D. Habersat, R.P. Sharma, T. Venkatesan, C.J. Scozzie, Matt Ervin, B.R. Geil, A.J. Lelis, S.J. Dixit, and R.S. Bhattacharya, AlN Thin Films Deposited by Pulsed Laser Ablation, Sputtering and Filtered Arc Technique, *Thin Solid Films*, **398-399**, 575 (2001).

64. J.R. Tuck, A.M. Korsunsky, D.G. Bhat, and S.J. Bull, Indentation Hardness Evaluation of Cathodic Arc Deposited Thin Hard Coatings, *Surf Coat Tech*, **139**, 63-74 (2001).

65. A.K. Rai and R.S. Bhattacharya, Development of Hard-Soft Coatings for Enhanced Machining Performance Employing Large Area Filtered Arc Deposition Technique (CT77), Phase II Final Report prepared for The Edison Materials Technology Center (EMTEC), UES, Inc., Dayton, Ohio USA (2006).

66. A.K. Rai and R.S. Bhattacharya, Large Area Filtered Arc Deposition of Carbon and Boron Based Hard Coatings, Final Report prepared for The U.S. Department of Energy (Contract No. DE-FG02-99ER82911), UES, Inc., Dayton, Ohio USA (2003).

67. C.Y. Chan, K.H. Lai, M.K. Fung, W.K. Wong, I. Bello, R.F. Huang, C.S. Lee, and S.T. Lee, Deposition and Properties of Tetrahedral Amorphous Carbon Films Prepared on Magnetic Hard Disks, *J Vac Sci Tech*, **A19**, 1606-1610 (2001).

ALUMINUM OXIDE AND SILICON NITRIDE THIN FILMS AS ANTI-CORROSION LAYERS

C. Qu[1], P. Li[1], J. Fan[1], D. Edwards[1], W. Schulze[1], G. Wynick[1], R. E. Miller[1], L. Lin[1], Q. Fang[2], K. Bryson[3], H. Rubin[4], and X. W. Wang[1]*

1. Alfred University, Alfred, NY 14802, USA
2. Oxford Instruments Plasma Technology, Yatton, Bristol BS49 4AP, UK
3. Duke University, Durham, NC27708, USA
4. Plasma-Therm, St. Petersburg, FL 33716, USA

ABSTRACT

To reduce the weight of a vehicle, magnesium-aluminum alloys may be utilized. However, when such alloys are in contact with carbon steel fasteners, electrochemical corrosion takes place. To reduce the corrosion, ceramic thin film coatings have been proposed as barrier layers. This study focused on the feasibility of utilizing an aluminum oxide coating or a silicon nitride coating as a barrier layer. The aluminum oxide coating was fabricated via an electron beam evaporation technique, and the silicon nitride coating was fabricated via a plasma deposition technique. The substrates were 1050 carbon steel plates. The anticorrosion behavior of each coating was evaluated via the electrochemical impedance spectroscopy. Thin film coating characteristics were further investigated by scanning electron microscopy and energy dispersive X-ray spectroscopy.

INTRODUCTION

In the automobile industry, in order to reduce energy consumption rate, light materials such as magnesium-aluminum (Mg-Al) alloy materials may be utilized to replace the corresponding carbon steel parts. When an Mg-Al alloy part is in direct contact with a carbon steel fastener, an electrochemical cell is formed. The electrochemical potential of the cell will drive a current through an electrical conduction path when such a path is provided. To our best knowledge, there is no existing method to satisfactorily stop or reduce the corrosion current flow. In searching for a possible solution, insulating ceramic materials were proposed as barrier layers in-between the Mg-Al alloy materials and the carbon steel fasteners. Specifically, silicon nitride and aluminum oxide were considered, since both materials have high electrical resistivity values and are relatively stable in high temperature and/or corrosive environment. As an example, silicon nitride thin films coated on metallic bond-pads with thicknesses between 55 nm and 500 nm were almost impervious to salt solution over 1.5 days[1]. In a previous work, silicon nitride thin films with thickness of approximately 850 nm were coated on flat 1050 carbon steel substrates via a plasma enhanced chemical vapor deposition technique (PECVD)[2]. Based on the electrochemical impedance spectroscopy (EIS) measurements, the impedance modulus of such a silicon nitride film was relatively high. In this work, to study the correlation between the film thickness and the impedance modulus, silicon nitride thin films with various thicknesses were fabricated with either the plasma assisted atomic layer deposition technique (ALD) or the PECVD techniques. As for the aluminum oxide materials, thick films with the thickness larger than 10 μm fabricated by a plasma spray technique were coated on turbine engine blades to form thermal-barriers[3]. In the previous study, aluminum oxide thin films were coated on the 1050 carbon steel substrates via an electron beam vapor deposition technique[2]. However, the impedance modulus of the aluminum oxide film was substantially lower than that of the silicon nitride film. In this study, several trial-and-error steps were taken to increase the impedance modulus.

*Corresponding Author: fwangx@alfred.edu

EXPERIMENTS

1.   Flat 1050 Carbon Steel Substrates

A ribbon of cold-rolled 1050 carbon steel material with the thickness of approximately 1.5 mm was provided by MNP Co. The ribbon was then cut into plates with several different sizes, ranging from 1 cm X 1 cm to 7.6 cm X 7.6 cm. Before the coating process, the substrates were ground sequentially by several grit size sandpapers, starting with 220 grits and ending with 1,000 grits. Thereafter, the substrates were polished with Buehler's diamond suspension media with the average sizes of 9 and 3 micrometers and aluminum oxide suspension medium with the average size of 0.5 micrometers. Then, the polished substrates were immediately immersed in the diluted Blue-Gold solution (5 volume %), which was purchased from Modern Chemical, Inc. With an ultrasonic cleaner, the substrates were cleaned three times, with 70% Isopropyl alcohol rinsing after each cleaning step. The cleaned substrates were finally dried with the compressed air. The polished substrates appeared to be shiny, however, still had scratch lines from the original-manufacturing process. There was no intention to completely remove the scratch lines, as the ultimate objective of the study is to coat the fasteners which may have grooves and trenches. Since the carbon steel is known for its inclusion impurities[4], an FEI Co. Quanta 200F environmental scanning electron microscope (ESEM) was utilized to characterize such impurities, along with an attached energy dispersive X-ray spectroscopy (EDS) module. In Figure 1, a scanning electron microscope (SEM) top view is provided. There are approximately 220 defects per mm^2. The sizes of the defects vary. In Figure 2, the EDS spectrum for the steel is illustrated, in which the dominant element is iron. In Figure 3, an "inclusion" defect area is illustrated in the middle of the SEM image. The shape of the defect appears to be round, with its "diameter" being approximately 10 micrometers. In Figure 4, the EDS spectrum of the defect area is illustrated. Besides iron, sulfur, calcium, manganese and other elements are revealed.

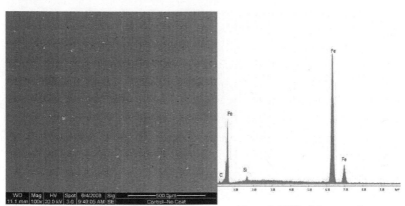

Figure 1. SEM top view of the control   Figure 2. EDS of the control

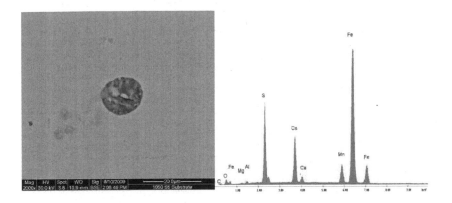

Figure 3. SEM top view of a defect area   Figure 4. EDS of a defect area

Impedance characteristics of an uncoated-polished 1050 carbon steel control sample were measured by Electrochemical Impedance Spectroscopy (EIS) measurements[5], with a wet cell area of 3 cm^2. After the surface area was exposed to the 5 wt % NaCl solution for 30 minutes, the input frequency of the power source was swept from a high frequency value to a low frequency value, with an AC excitation voltage of 10 mV. At each given frequency, the impedance was measured. In Figure 5, the impedance modulus is plotted as a function of the frequency for a control sample (bottom-red curve). As the frequency decreases from 10 KHz to 0.01 Hz, the impedance modulus value increases from 41 Ohm-cm^2 to 3.7Kilo-Ohm-cm^2. This result indicates that polished 1050 carbon steel itself cannot resist corrosion. For the anti-corrosion applications to be considered in this study, the main focus will be the impedance modulus values at 0.01 Hz. Visually, corrosion was observed after the 30 minute soaking with the salt water.

2.   Aluminum Oxide Coatings on 1050 Carbon Steel Substrates
Aluminum oxide thin films were deposited via the electron beam evaporation technique[2]. A film was deposited at a substrate temperature of approximately 190 °C, with a deposition rate of approximately 1 Å/s and a film thickness of approximately 200 nm (Film A1). In Figures 6 and 7, two SEM top views with two different magnifications are illustrated. In comparison with the uncoated substrate illustrated in Figure 1, the number of the defects per unit area is reduced from approximately 220 defects per mm^2 to approximately 20 defects per mm^2, as shown in Figure 6. That is, compared with the substrate before the coating, the majority of the steel surface is covered by the aluminum oxide coating. In Figure 7, a partially covered defect area is illustrated. In Figure 5, the impedance modulus of this film is plotted as a function of the frequency (middle-black curve). In comparison with the impedance modulus curve for the uncoated 1050 carbon steel substrate (bottom-red curve), the impedance modulus of the coated sample (Film A1) is generally higher than that of the control, at a given frequency between 1 KHz and 0.01 Hz. In particular, at 0.01 Hz, the impedance modulus of the coated sample is approximately 7.1×10^3 ohm-cm^2, while that of the control is 3.7×10^3 ohm-cm^2. With the longer deposition time, another film was fabricated with the film thickness of approximately 700 nm (Film A2). In Figure 5, the top-blue curve is for Film A2. At 0.01 Hz, the impedance modulus value is approximately 3.6×10^4 Ohm-cm^2, which is higher than that of 200 nm film (Film A2).

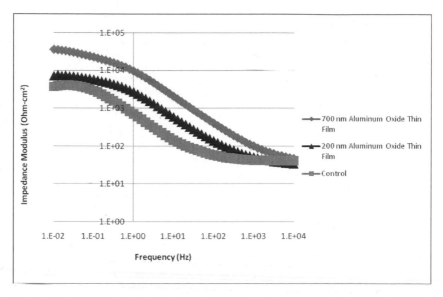

Figure 5. Impedance modulus is plotted as the function of the frequency. The bottom-red curve is for the uncoated 1050 steel substrate. The middle-black curve is for single layer aluminum oxide film with the thickness of approximately 200 nm (Film A1). The top-blue curve is for single layer aluminum oxide film with the thickness of approximately 700 nm (Film A2).

Figure 6. SEM top view with magnification of 200 for an aluminum oxide film with the thickness of approximately 200 nm (Film A1)

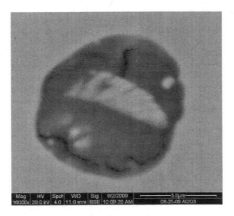

Figure 7. SEM top view of Film A1, with magnification of 10K

In a furnace with the ambient atmosphere, the post deposition heat-treatment was conducted for a film fabricated with the same deposition conditions as Film A1, with the film thickness of approximately 150 nm. The heat-treatment temperature for this sample (Film A3) was approximately 350 °C, and the heat-treatment time duration was approximately 40 min. At a given frequency, the impedance modulus of Film A3 is always higher than that of Film A1. Specifically, at 0.01 Hz, the impedance modulus of Film A3 is 2.6×10^4 Ohm-cm^2, which is higher than that of Film A1 (7.1×10^3 Ohm-cm^2). For another film fabricated under the same condition as Film A3, after the heat-treatment, another layer of aluminum oxide film with the thickness of approximately 120 nm was deposited. The total thickness of the coating after the second deposition is approximately 270 nm (Film A4). In Figure 8, the impedance modulus plot for Film A4 is illustrated (green curve). At 0.01 Hz, the impedance modulus value for Film A4 is approximately 4.6×10^5 Ohm-cm^2, which is about 60 times larger than that of Film A1. A layer of UV-curable materials containing cerium oxide material was applied on top of the bare 1050 steel substrate, Film UV[6]. In Figure 8, the dark-red curve is the impedance modulus plot for Film UV. At 0.01 Hz, the impedance modulus value for Film UV is approximately 1.6×10^5 Ohm-cm^2. In addition, a layer of the UV curable materials was applied onto a film fabricated under the same conditions as Film A4, Film UV-A4. In Figure 8, the blue-purple curve is the impedance modulus plot for Film UV-A4. At 0.01 Hz, the impedance modulus value of Film UV-A4 is approximately 2.1×10^6 Ohm-cm^2.

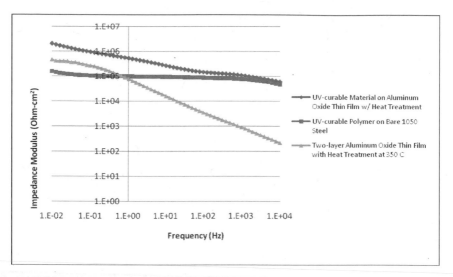

Figure 8. Impedance modulus is plotted as a function of the frequency. The green curve is for Film A4. The dark-red curve is for Film UV. The blue-purple curve is for Film UV-A4.

3.   Silicon Nitride Coatings on 1050 Carbon Steel Substrates

Silicon nitride thin films were either coated by the plasma assisted atomic layer deposition (ALD) technique or the plasma enhanced chemical vapor deposition (PECVD) techniques[7]. The substrate temperatures were 250 - 300 °C. The precursors for plasma ALD included $SiH[N(CH_3)_2]_3$ (silane), nitrogen gas, and hydrogen gas. The precursors for PECVD included silane (100% or 3% in argon), ammonia, and/or nitrogen gas. There were two film thickness groups. The thickness for the first group is 20 - 60 nm, and the thickness for the second group is 300-850 nm. In Figure 9, an SEM top view is provided for a typical film in the first group, with the film thickness of approximately 60 nm. As illustrated in the figure, some of the substrate features are still visible. In Figure 10, another SEM top view is illustrated for a partially covered defect area, in which the 60 nm film coverage is incomplete. In Figure 11, the red curve is the impedance modulus plot for 20 nm silicon nitride film, the purple curve is for 40 nm silicon nitride film, and the moss-green curve is for 60 nm silicon nitride film. At 0.01 Hz, the impedance modulus value for the 60 nm film is approximately 6.1 X10^4 Ohm-cm^2, which is higher than both impedance modulus values for the 40 nm and 20 nm films (2.1-2.2 X 10^4 Ohm-cm^2). In Figures 12-13, two SEM top views are illustrated for a silicon nitride film with a thickness of approximately 300 nm. In Figure 12, with the magnification of 100K, most of the surface fine features are covered by the film. However, in Figure 13, with the magnification of 500, the inclusion defect in the middle of the image cannot be completely covered by the film. In Figure 11, the black curve is for the impedance modulus plot of 300 nm silicon nitride film. At 0.01 Hz, the impedance modulus is approximately 2.1 X 10^5 Ohm-cm^2. In Figures 14 and 15, two SEM top views are illustrated with two different magnifications (20K and 2K), for a film with the thickness of approximately 600 nm. In Figure 14, most of the boundaries between two adjacent grains are "sealed."

However, in Figure 15, the coverage over each of the inclusion defect areas is not complete. In Figure 11, the orange curve is for the impedance modulus plot of 600 nm silicon nitride film. At 0.01 Hz, the impedance modulus is 4.1×10^5 Ohm-cm^2. In Figures 16 and 17, two SEM top views with two different magnifications are illustrated, for a film with the thickness of approximately 850 nm. In Figure 16, the boundaries between two adjacent grains are "sealed." In Figure 17, the defects of the substrate are covered by the coating. In Figure 11, the top-blue curve is the impedance modulus plot for 850 nm silicon nitride film. The impedance modulus at 0.01 Hz is approximately 1.4×10^6 Ohm-cm^2. In Figure 18, both impedance modulus curves for Film UV-A4 and 850 nm silicon nitride Coating are illustrated. At 0.01 Hz, the impedance modulus values for these two films are very close. However, to reach the impedance modulus value of approximately 2×10^6 Ohm-cm^2, the total number of steps to fabricate each film is different. For Film UV-A4, the toal number of steps is four. However, there is only one step for 850 nm silicon nitride film fabrication.

Figure 9. SEM top view of silicon nitride coating on 1050 carbon steel, with a thickness of 60 nm

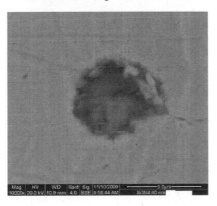

Figure 10. SEM top view of a partially covered inclusion defect area

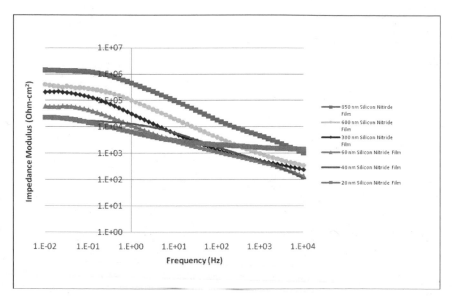

Figure 11. Impedance modulus is plotted as a function of the frequency. The red curve is for 20 nm silicon nitride film, the purple curve is for 40 nm silicon nitride film, the moss-green curve is for 60 nm silicon nitride film, the black curve is for 300 nm silicon nitride film, the orange curve is for 600 nm film, and the top-blue curve is for 850 nm film.

Figure 12. SEM top view of 300 nm silicon nitride film on 1050 carbon steel

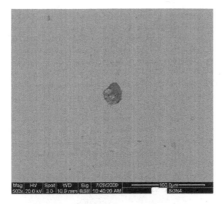

Figure 13. SEM top view of 300 nm silicon nitride film which cannot cover a defect area

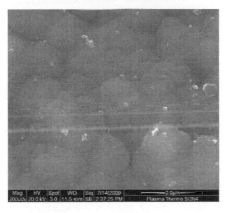

Figure 14. SEM top view with magnification of 20 K, for 600 nm silicon nitride film

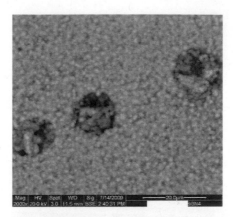

Figure 15. SEM top view with magnification of 2K, for 600 nm silicon nitride film

Figure 16. SEM top view with magnification of 100K, for 850 nm silicon nitride film

Figure 17. SEM top view with magnification of 10K, for 850 nm silicon nitride film

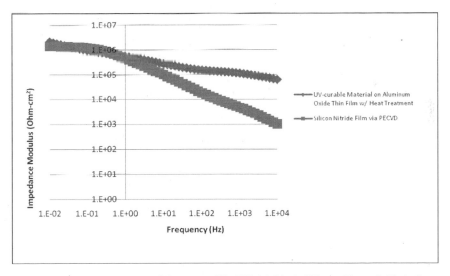

Figure 18.   Impedance modulu curves: Film UV-A4 (blue), 850 nm silicon nitride (red)

CONCLUSIONS AND DISCUSSIONS

For the two-layer aluminum oxide film fabricated with three-steps, the highest impedance modulus value at 0.01 Hz is approximately 4.6×10^5 Ohm-cm^2. With an additional UV curable cerium oxide layer, the impedance modulus value is increased to approximately 2.1×10^6 Ohm-cm^2. For 850 nm silicon nitride film fabricated with one-step, the impedance modulus value is 1.4×10^6 Ohm-cm^2. In this paper, the main concern is the coverage over the 1050 carbon steel substrate. Currently, we are examining other factors which may also affect the impedance modulus values, such as the film densities. Besides flat 1050 carbon steel substrates, aluminum oxide and/or silicon nitride thin films were coated onto M10 carbon steel bolts. On some of the coated bolt's heads, UV curable materials were then coated as the cap-layer. Based on the preliminary corrosion testing results, the combination between the silicon nitride thin film and the UV curable cap-layer seemed to perform better than other coating combinations.

REFERENCES
1.   M. Jo, and B. Noh, "A study on the feasibility of silicon nitride thin film as diffusion barriers over IC chip packaging," J. Ind. Eng. Chem. Vol. 8, pp. 458-463, 2002.
2.   Y. Liu, C. Qu, R. E. Miller, D. D. Edwards, J. H. Fan, P. Li, E. Pierce, A. Geleil, G. Wynick, and X. W. Wang, " Comparison of Oxide and Nitride Thin Films – Electrochemical Impedance Measurements and Materials Properties," to be published in the Proceedings of 8th Pacific Rim Conference on Ceramics and Glass Technology, Vancouver, BC, Canada, 2009. To appear in Ceramic Transactions, Volume 213, Nanostructured Materials and Systems, Edited by Sanjay Mathur.
3.   C. Li, G. Yang, and A. Ohmori, "Relationship between particle erosion and lamellar microstructure for plasma-sprayed alumina coatings, Wear, Vol. 260, pp. 1166-1172, 2006.
4.   M. Fernandes, N. Cheung, and A. Garcia, "Investigation of Nonmetallic Inclusions in Continuously Cast Carbon Steel by Dissolution of the Ferritic Matrix," *Mater. Charact.,* Vol. 48, pp. 255-261, 2002.
5.   Solartron potentiostat/galvanostat 1287 and Solartron 1260 frequency response analyzer (FRA)
6.   The UV curable materials were purchased from Chemat Technology, Northridge, CA 91324. There were two parts: CeO_2 precursor ($Ce(OCH_2CH_2OCH_3)_3$) and reaction initiator (AR Base).
7.   The ALD system is the Oxford Instruments FlexAL system. The PECVD systems include the Trikon/Aviza Planar FxP system, Plasma-Therm 790+ system and the Advanced Vacuum Vision 310 system.

DISCLAIMER

This material is based upon work supported by the Department of Energy National Energy Technology Laboratory under Award Number DE-FC26-02OR22910. This report was prepared as an account of work sponsored by an agency of the United States Government. Neither the United States Government nor any agency thereof, nor any of their employees, makes any warranty, express or implied, or assumes any legal liability or responsibility for the accuracy, completeness, or usefulness of any information, apparatus, product, or process disclosed, or represents that its use would not infringe privately owned rights. Reference herein to any specific commercial product, process, or service by trade name, trademark, manufacturer, or otherwise does not necessarily constitute or imply its endorsement, recommendation, or favoring by the United States Government or any agency thereof. The views and opinions of authors expressed herein do not necessarily state or reflect those of the United States Government or any agency thereof. The partial financial support from US CAR program is acknowledged.

PLASMA NITRIDED AUSTENITIC STAINLESS STEEL FOR BIOMEDICAL APPLICATIONS

O. Gokcekaya, S. Yilmaz, C. Ergun, B. Kaya, O. Yücel

Istanbul Technical University, Mechanical Engineering, 34437, Istanbul, TURKEY.

Istanbul Technical University, Adnan Tek. Research Center, 34469, Istanbul, TURKEY.

Abstract

Austenitic stainless steels are widely used as a non-implantable medical device because of good corrosion resistance and moderate strength. However, they have low surface hardness and wear resistance properties. In order to improve mechanical, chemical and biological properties, surface modifications have been applied to these biomaterials. In this study, AISI 316 austenitic stainless steels were plasma nitrided under different process parameters including temperature (450, 500, 550 ^0C) and time (2, 4 and 9 hr) at a gas mixture of 25%N_2–75% H_2. The nitrided surfaces were characterized by X-ray diffraction (XRD). MicroVickers hardness tests were done on the cross sections and the surface of the nitrided specimens to investigate the hardness profile. The important effects of nitriding temperature and time on the microstructure and hardness value of nitrided surface layers were introduced.

Introduction

Metallic materials, such as stainless steel, cobalt-chromium alloys, commercially pure titanium, and titanium alloys have been used as an implant/prosthesis materials in many different forms in load bearing biomedical applications, such as dental implants and restorations, bone and joint replacement, as well as heart valves, etc. [1-4]. Stainless steel was one of the first successfully used implant material in the surgical field, as shown in Figure 1 [2].

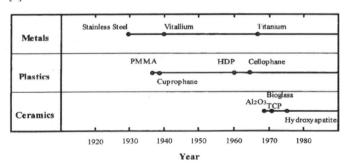

Fig. 1—History of metals, plastics, and ceramics for biomedical applications [2].

Metals corrode in contact with aggressive body fluids or tissue [6]. In general, when a biometal is implanted into a human body, a sudden reaction between implant surfaces and neighboring tissues occurs, which basically determines the response of the tissue against the material in the following periods. Therefore, the surface properties become critically important which will determine the in vivo performance of that particular biomaterial [7-8].

135

Various surface modifications techniques can be applied to increase the performance of implant/prosthesis in order to meet the requirements demanded for a wide variety of clinical applications. The main objective in these techniques is to tailor the surface properties by changing the composition and micro structure, while leaving the bulk mechanical properties of the material intact [11-14]. For austenitic stainless steels, the improvement of surface hardness and wear resistance was accompanied by reduction of its corrosion resistance after nitriding at temperature above 450 C due to chromium nitride precipitation and depletion of chromium in substrate [9-10].

Any effort to further improve in vivo chemical and biological properties associated with the mechanical properties, such as wear resistance of the implants, are greatly appreciated by the implant/prosthesis community.

In recent years, more and more research on developing new technologies used for nitriding of austenitic stainless steels has been performed. These technologies include plasma nitriding, plasma immersion implantation, ECR ion nitriding, rf plasma nitriding, low-pressure plasma assisted nitriding and high current-density ion beam nitriding [12–13]. Plasma assisted thermo-mechanical treatment is one of the popular surface modification technique applied on metallic materials [11-14].

The aim of the current study is to understand the effects of ion nitriding time and temperature on the microstructure, morphology and micro-hardness properties of 316L Stainless Steel for orthopedic applications.

Experimental Details

Stainless steel was used in the present experiment with a chemical composition of 0.022wt% C, 0.79wt% Si, 1.6wt% Mn, 0.25 wt% P, 0.002wt% S, 15.30wt% Cr 14.09wt% Ni, 2.63 wt% Mo and 0.05wt% Cu, respectively. The specimens were cut from cylindrical bars with diameters of 8 mm and thickness of 15 mm. In the first, they were grinded starting from coarser grinding papers (320) to finer grinding paper (1200) and mechanically polished to a near mirror finish. In the second the specimens were cleaned by trichloroethylene to free any kind of oil and dust before nitriding procedure.

A home-made DC plasma-nitriding equipment was used to nitride the specimens. A high silica tube with a 200 mm in diameter, 400 mm in height was used as the nitriding chamber (Figure 2). Cathode was placed in the middle of the chamber.

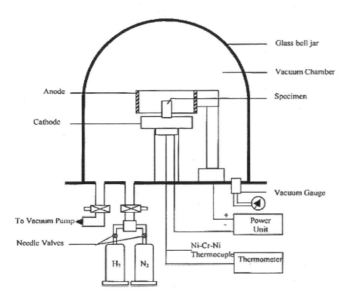

Fig. 2. Schematic diagram of the ion nitriding system [17].

A stainless steel disc with a 30 mm in diameter which was positioned to the middle of the chamber was used as a cathode. The specimens were placed on to this cathode plate. A disc-shaped metal plate with a 130 mm in diameter and 2.5 mm in thickness was used as an anode. A thermocouple was connected to the specimen thorough the cathode to monitor the specimens temperature.

The tube was first evacuated to 10 Torr with a single stage rotary vacuum pump maintained vacuum pressure. The same vacuum pump was also used to maintain the desired vacuum level throughout the nitriding processes.

Prior to the plasma nitriding, the specimens were subjected to cleaning by hydrogen sputtering for 30min to remove surface oxides and other contaminates. Then, the plasma nitriding was performed in a gas mixture of $80\%N_2$–$20\%H_2$, with a total pressure of 1 kPa. The process temperatures were 450, 500 and 550 °C, the process times were 2, 4 and 9 h.

The phases of the nitrided layers were analyzed by X-ray Diffraction (Panalytical type X-Pert Pro diffractometer). Cu-Kα radiation (λ = 1.5406 nm) produced at 45 kV and 40 mA scanned the diffraction angles (2θ) between 20° and 70° at every 0.145° for 1 s. Diffraction signal intensity throughout the scan was monitored and processed with X-Pert HighScore Plus software.

Vickers hardness tests were performed on the cross-sections and on the surface of specimens using a Shimadzu HMV Micro Hardness Tester. The indentation load was 100 g and the indentation time was 10 s.

Results and Discussion

1. Materials Characterization

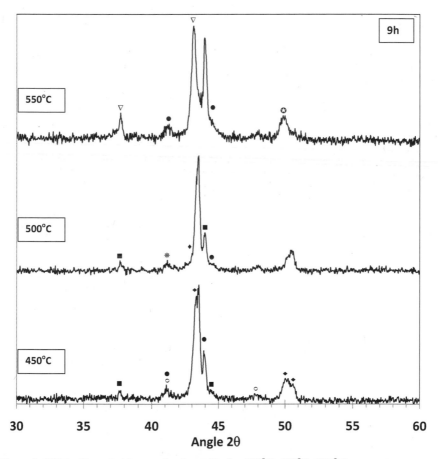

Figure 3: XRD of ion nitride samples for at 9hr for 450°C, 500°C, 550°C.

( ) Fe, (●)Fe₃N, (o)Fe₄N, (♦)FeN0.076, (■)CrN, (✳) Fe3Ni N, (∇)Cr₂N, (✿)FcN0.093

XRD graphs of nitrided 316L stainless steel at the different process parameters are given in Figure 3. The microstructure of the untreated 316L completely consisted of fcc-γ phase. After nitriding process, a combination of CrN, Fe₃N, Cr₂N, Fe₄N, FeN₀.₀₇₆, FeN₀.₀₉₃ phases were formed depending on the nitriding temperature.

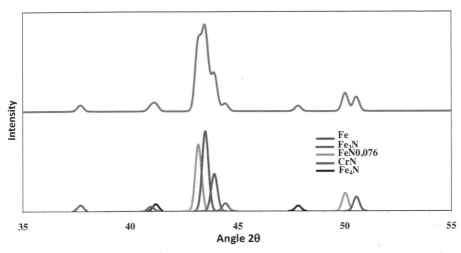

Figure 4: Detailed phase analysis in XRD pattern for 316L stainless steel ion nitride samples for 9hr at 450°C.

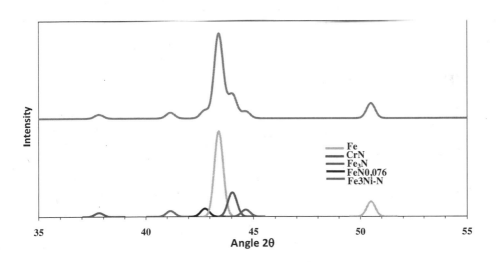

Figure 5: Detailed phase analysis in XRD pattern for 316L stainless steel ion nitride samples for 9hr at 500°C.

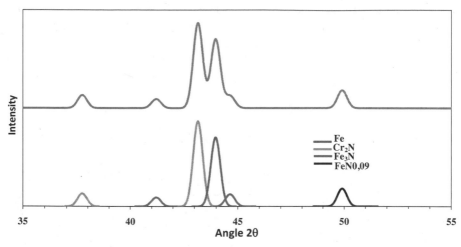

Figure 6: Detailed phase analysis in XRD pattern for 316L stainless steel ion nitride samples for 9hr at 550°C.

Increasing the temperature from 450 to 550°C, Fe_4N phase disappeared at 41° and 48° angles. The intensities of CrN phase decreased as the temperature increased, and disappeared at 550°C on 43° angle. Similarly, CrN disappeared at 500°C with increasing temperature on 50° angle, and Cr_2N phase increased as the temperature increased. With increasing temperature free Fe atoms may bond with N atoms then formed Fe_3N, Fe_4N, FeN0,07, FeN0,09. Similarly, CrN and Cr_2N phases were formed with the increasing temperature. This can be seen more clearly in the detailed phase analysis in XRD pattern shown in Fig. 4,5,6 respectively.

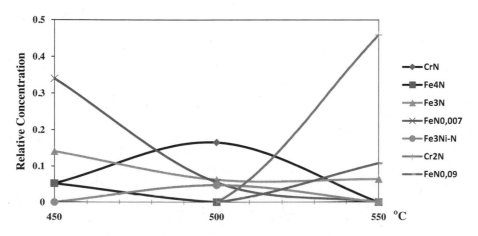

Figure 7:Relative amounts of different phases for ion nitrided stainless steel at different temperatures (450°C, 500°C, 550°C)

The relative XRD peak heights were plotted as a function of ion nitriding temperatures in Fig7. The relative peak heights show the changes in the amounts of a particular phase, but are not normalized to give absolute concentrations. The most hardness of surface was obtained at 450°C nitriding temperature because of Fe_3N, Fe_4N, $FeN0,07$ phases, as seen in Fig7. Increasing nitriding temperature diffusion occurred and also case depth increased. But surface hardness of the nitrided samples decreased with increasing temperature. The reason of this may be the occurrence of CrN and Cr_2N phases.

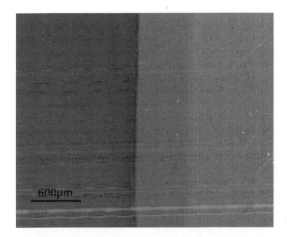

Figure 8: Surface of ion nitrided sample for 9hr at 550°C.

Figure 9: Surface of ion nitrided sample for 9hr at 500°C.

Figure 10: Surface of ion nitrided sample for 9hr at 450°C.

The optical micrographs of ion nitrided samples at 450°C, 500°C, 550°C for 9 hr were shown in Figure 8,9,10 respectively. The micro-hardness measurements were made on the cross section with MicroVickers hardness test.

2. Hardness Measurements

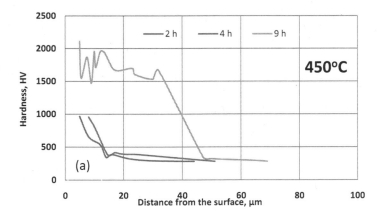

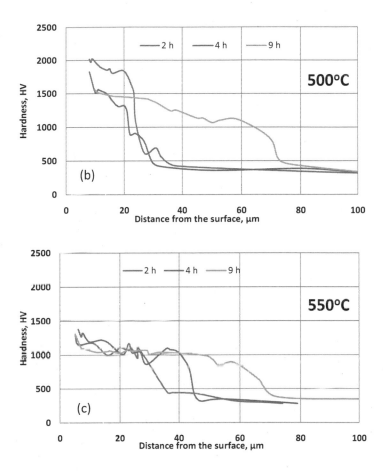

Figure 11: Microhardness profile of 316L stainless steel after ion nitriding a) 450°C b) 500°C c) 550°C

Microhardness profiles of the specimens nitrided at different process conditions are given in Figure 11,12. The initial hardness of stainless steel was 280 HV. There was an increase in the hardness after nitriding that is caused by the increase of the nitrogen concentration and the formation of new phases like CrN, Fe_3N, Fe_4N on the surface of the material. The hardness values didn't decrease through the diffusion zone to approach the core of the sample, there are fluctuations because of CrN, Cr_2N are formed at different case depths. With the increase in the temperature, the diffusion highly activated and the nitride formation and gas saturation were accelerated, which resulted growth in the diffusion layer, but a few decreasing surface hardness because of decreasing nitrogen concentration

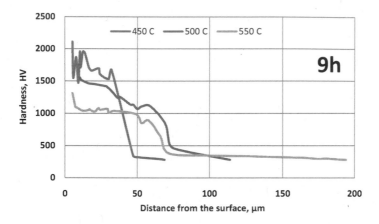

Figure 12: Microhardness profile of 316L stainless steel after ion nitriding for 9 hours

As seen Fig12, with increasing the temperature, case depth increased. On the other hand surface hardness decreased. The reason of decreasing the hardness can be decreasing the amount of Fe_3N, Fe_4N, $FeN0,07$ phases. This results can be seen in Fig7.

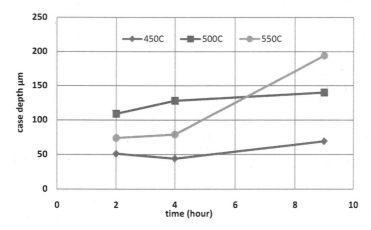

Figure 13: The effect of process temperature on the case depth of nitrided samples

Process temperature and time affected the case depth of nitrided samples are shown in Figure 13. Process time was more effective than temperature in the case depth of the nitrided samples for 550°C. The largest case depth was measured as 194 μm for 9 hour 550 °C.

Conclusion

In this research, 316L stainless steel has been nitrided by using ion nitriding process. CrN, Cr_2N, FeN, Fe_3N and Fe_4N phases formed on the surface of specimen after nitriding as shown Fig3,4,5,6. The microhardness and compound layer thickness increased with increasing nitriding time and temperature for some cases. The surface microhardness of 316L stainless steel for 9 hour at 450°C nitriding conditions reached to a value of 2111 HV was approximately 7,5 times higher than untreated materials as seen Fig11(a). The nitriding time affected case depth strongly at 550°CFig11(c). The case depth values of nitrided specimens for 9 hour procedure time were 3.5-4 times higher than nitrided specimens for 2 hour at 550°Cas seen Fig11. These results showed us, the most hardness surface obtained at 450°C and the largest case depth could be obtained with increasing the ion nitriding temperature.

Acknowledgments

This work is founded by TUBITAK (Project # 106M202).

References

[1] J. Black, G. Hastings, Handbook of Biomaterial Properties, p.167, Chapman&Hall,

[2] J.B. Park: in *The Biomedical Engineering Handbook Volume 1*, J.D. Bronzino, ed., CRC Press LLC, Boca Raton, FL, 2000, pp. IV-1-IV-8.

[3] Park JB, Lakes RS. Biomaterials an introduction, 2nd ed. NY: Plenum Press, 1992.

[4] T.S. Hin, Engineering Materials for Biomedical Applications, World Scientific, Singapore (2004).

[5] Bolton W. Engineering and material technology. 2nd ed. Boston: Newnes, 1995. p. 217.

[6] Neumann HG, Beck U, Drawe M, Steinbach J. Surf Coat Technol1998;98:1157 .

[7] R. Wei, T. Booker, C. Rincon, J. Arps, Surf. Coat. Technol., 186, 305 (2004).

[8] O. Ozturk, U. Turkan, A.E. Eroglu, Surf. Coat. Technol., 200, 5687 (2006).

[9] A. Ramchandani, J.K. Dennis, Heat Treat. Met. 2 (1988) 34.

[10] K. Marchev, C.V. Cooper, J.T. Blucher, B.C. Giessen, Surf. Coat. Technol. 99 (1998) 225.

[11] A. Molinari, G. Straffelini, B. Tesi, T. Bacci, G. Pradelli, Wear, 203, 447 (1997).

[12] S.-P. Hannula, P. Nenonen, J.-P. Hirvonen, Thin Solid Films 181 (1989) 343.

[13] P.Smith, R.A. Buchanan, J.R. Roth, S.G. Kamath, J. Vac. Sci. Technol. B 12 (1994) 940;

[14] S.L.R. Silva, L.O. Kerber, L. Amaral, C.A. Santos, Surf. Coat. Technol., 16, 342 (1999).

[15] A. Çelik, O. Bayrak, A. Alsaran, I. Kaymaz, A.F. Yetim, Surf. Coat. Technol., 202, 2433 (2008).

[16] R. Wei, T. Booker, C. Rincon, J. Arps, Surf. Coat. Technol., 186, 305 (2004).

[17] A. Arslan, A. Çelik, Materials Characterization 47 (2001) 207– 213

A Report on the Microstructure of As-Fabricated, Heat Treated and Irradiated ZrC Coated Surrogate TRISO Particles

Gokul Vasudevamurthy † §, Yutai Katoh §, Jun Aihara‡, Shohei Ueta‡,
Lance L. Snead §, Kazuhiro Sawa‡

§ Materials Science & Technology Division, Oak Ridge National Laboratory, TN, USA
‡ High Temperature Fuel & Materials Group, Japan Atomic Energy Agency, Ibaraki-ken, Japan
† University of Tennessee - Knoxville, TN, USA

Abstract

Zirconium carbide is a candidate to either replace or supplement silicon carbide as a coating material in fuel particles for high temperature gas-cooled reactor fuels. Desirable characteristics of ZrC as a fuel coating include high melting point, adequate fission product retention capability, appropriate neutronic characteristics, resistance to fission product palladium corrosion, and reasonable thermal conductivity. However, there is not sufficient data to demonstrate the suitability of ZrC for nuclear fuel applications. The US and Japan have initiated a collaborative Study to evaluate the feasibility of using ZrC as a fuel coating material, in which microstructural, thermophysical, and thermomechanical properties of developmental ZrC coatings are being evaluated for both unirradiated and irradiated conditions. This paper presents the results of a joint endeavor to study the microstructural evolution, including the effects of C/Zr ratio, in as-fabricated and heat treated samples of nominally stoichiometric and carbon rich ZrC coated surrogate microspheres irradiated to fluence of 2 and 6 dpa at 800 and 1250°C. Grain growth was the primary irradiation effect observed in all the samples. Microstructural examination revealed greater grain growth in stoichiometric ZrC coatings predicting mechanical unsuitability of currently proposed ZrC for nuclear fuel coating applications.

1.0 Introduction

High temperature gas cooled reactors (HTGR) are promoted to be the next generation nuclear power plants for both electricity and process heat applications, such as hydrogen production or petroleum refinement. HTGRs offer attractive features such as high burn up (~ 200 GWd/MTU) and high output temperature, but demand fuel and structural materials which perform satisfactorily without sustaining damage at both operating (_1400°C) and accidental transient conditions (>1600°C). To satisfy these conditions, HTGRs will use TRISO (Tristructural Isotropic) fuel particles embedded in an inert matrix material such as graphite. A typical TRISO fuel particle (Figure 1) consists of an inner uranium oxide or oxycarbide fuel kernel surrounded by four functional layers. The first layer consists of a 50% dense carbon (Buffer) layer to assist as a buffer plenum to store gases produced during the fission reaction and accommodate fuel swelling during operation. The second layer is a dense pyrolytic carbon (IPyC), which acts as a secondary fission product containment layer and provides a substrate for the primary containment layer. The third layer functions as the primary containment vessel by providing the necessary mechanical strength to the entire fuel particle and presenting a barrier to radiotoxic fission products. This layer currently consists of chemical vapor deposited silicon carbide (CVD SiC). The fourth and outermost layer is again made of dense pyrolytic carbon (OPyC) and functions as a cushion layer to protect the SiC layer during compact fabrication and handling. Of these three layers the integrity of the ceramic layer, which serves as the primary fission product barrier, is crucial and hence the mechanical integrity of this layer under both operating and transient temperatures must be assured. Required characteristics of this layer include: high melting point, tolerance of neutron irradiation, resistance to fission product corrosion, good thermal conductivity, appropriate neutronic properties, high temperature chemical stability, and adequate mechanical strength[1]. Although the currently used SiC satisfies most of these characteristics, it is expected to suffer from certain important disadvantages such as susceptibility to corrosion by fission product palladium[2, 3]. This corrosion problem will increase in deep burn plutonium based fuels where fission product palladium production is more significant. Zirconium carbide (ZrC) has demonstrated a good potential to replace SiC as a coating material. ZrC has shown good resistance to fission product containment capabilities[2] and palladium corrosion[3] comparable to SiC, with the possibility of higher operating temperature limits.

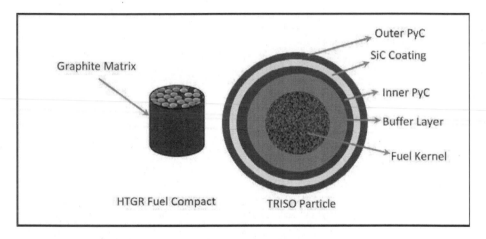

Fig. 1. Typical TRISO fuel particle and fuel compact

Although ZrC has often been discussed as a performance enhancing coating layer, very little data is currently available on the thermophysical and thermomechanical properties of ZrC in the desired high temperature and irradiation fluence ranges encountered in HTGR cores. The ability of ZrC to exist in different stiochiometric compositions further complicates the matter, because the variation of C/Zr ratio is expected to significantly alter the high temperature properties. Additionally, very limited authoritative data is available on the irradiation behavior of ZrC. Experiments conducted at the Oak Ridge National Laboratory in the 1970s indicated promising irradiation performance of ZrC coated TRISO fuel and hot pressed ZrC pellets [5-7]. However, other than macroscopic observation of coating integrity, no detailed results of the microstructural evolution or the mechanical properties were available from these experiments. Keilholtz et. al. [5] reported the absence of grain boundary separations at low fluence (<1dpa) and temperatures (300-700°C) in hot pressed ZrC. Recent experiments with proton[8, 9], neutron[10] and ion[11] irradiation of hot pressed and zone refined ZrC has shown no drastic damage to the microstructure. There was no amorphization and void formation, but defect loops (probably Frank loops [10]) were observed with increasing fluence and temperatures. Overall, the results of the limited experiments reported in the literature have established that ZrC is microstructurally stable under the tested conditions. The stability has been attributed in part to the strong covalent bonded crystal structure of ZrC and the presence of a high density of carbon vacancies [8, 9, 11]. However no models explaining the reported stability have been proposed. Based on the limited yet encouraging historical observations, a majority of nuclear fuel researchers opine that further detailed investigations are required to examine if ZrC is a viable replacement for, or supplemental layer in addition to SiC in fuel coating applications. Recently, both the United States and Japan have renewed their interests in evaluating the suitability of ZrC under their respective HTGR fuel development programs. The investigations planned pertain to irradiation behavior and mechanical properties evaluation of ZrC, especially with varying stoichiometry. The main objective of this study is the comparative microstructural observations of irradiated carbon rich (C/Zr~1.4) and near-stoichiometric ZrC.

2.0 Experiment

The samples described in this paper are surrogate yittria stabilized zirconia (YSZ) particles coated with ZrC by the Japan Atomic Energy Agency (JAEA) employing a zirconium bromide fed CVD process. The details of fabrication are discussed elsewhere [12, 13, 14]. Six different samples were irradiated in the High Flux Isotope Reactor (HFIR) at the Oak Ridge National Laboratory (ORNL) and four of those are presented and discussed in this paper. The details of the irradiated samples including the stoichiometry (as measured and reported by JAEA), and the irradiation conditions are shown in Table 1. The irradiation conditions were chosen to simulate the operating conditions of the fuel in a HTGR. The neutron dose values of 2 and 6 dpa represented of the half and quarter life time dose of HTGR fuels whereas 800°C and 1250°C were indicative of lower and upper limits of fuel operating temperature. The C/Zr ratios were determined by combustion analysis at JAEA. Samples 3002HT, 2048HT, and 2003HT shown in Table 1 were heat treated at 1800°C for 1h prior to irradiation. The purpose of this heat treatment was to simulate the effect of high temperatures experienced by actual TRISO particles during the pelletization process. The samples after irradiation were extracted from the capsules at the Irradiated Materials Examination Facility (IMET) at ORNL and shipped to the Low Activation Materials Development and Analysis Laboratory (LAMDA) for further examinations. Intact samples (with unbroken coatings) were cleaned and mounted in epoxy resin, ground to the mid plane, and polished using diamond compounds. Identical grinding steps were employed during preparation of the unirradiated samples. Microstructural observations were performed using a Phillips XL-30 and a JEOL 6500F field emission scanning electron microscope (SEM). It is noted here that the sample lot number and the sample names have been used interchangeably while indexing the figures and the following discussions.

Table 1: Sample Matrix and Experimental conditions

Sample Lot Number	C/Zr*	ZrC Thickness* (mm)	OPyC Thickness* (mm)	IPyC Thickness* (mm)	Heat Treated (at 1800°C)	Fluence (dpa)¥	Irradiation Temperature* (°C)
3002	1.0	21	30	37	No	2	800, 1250
						6	1250
3002HT	1.0	21	30	37	Yes	2	800, 1250
						6	1250
2048HT	1.0	27	0	34	Yes	2	800,1250
						6	1250
2003HT	1.4	28	0	33	Yes	2	800, 1250
						6	1250

Indicates nominal values, ¥ 1 dpa (displacement per atom) per 1×10^{25} neutrons/m² (E > 0.1 MeV)

3.0 Results and Discussion
3.1 Unirradiated Microstructure

Figure 2. shows the representative pre-irradiation microstructure of the as-fabricated and heat treated particles that were received from JAEA. Sample 3002 showed very small grains (< 100 nm) with no layered structure. The grains were homogenous and the microstructure showed some pores. Sample 3002HT, produced from sample 3002 with an additional heat treatment, showed the presence of large grains (>2-3 μm) in addition to some pores. Sample 3002HT also did not show any layered structure.

A comparison of samples 3002HT and 2048HT, both heat treated and with nominal stoichiometry of 1.0, showed a significantly different microstructure. The grains in sample 3002HT did not exhibit any particular pattern whereas sample 2048HT clearly showed a two region layered structure with annular grain bands. The region close to the IPyC contained finer grains compared to the region close to the outer periphery. This distinctive structure could be attributed in part to change in experimental conditions during a particular coating run resulting in layers with different carbon compositions. However, no such changes were confirmed during the process. The predecessor of sample 2048HT (not irradiated in this experiment, Fig. 3) shows a clear layered structure which had been retained during additional heat treatment. Post-heat treatment, the inner region with finer grains had undergone less grain growth compared to the outer layer. This anomaly could be attributed to the presence of carbonaceous interlayer in the inner region which could not be analytically distinguished due to limitations in resolution of back scattered images and electron dispersive spectroscopy. This hypothesis was further confirmed from the microstructural observations of the carbon rich sample 2003HT (Fig.2). The microstructure revealed a ring structure with alternating dark and bright annular bands. The dark bands represented the excess carbon deposited between the coating layers. The presence of a carbon interlayer was also partly confirmed by Aihara et.al. [12, 13, 14,] in reporting the transmission electron microscope observations of the heat treated samples. The presence of the carbon interlayer seemingly inhibited grain growth in sample 2003HT during heat treatment. Extending the observations, it could be concluded that though samples 3002HT and 2048HT were expected to have a nominal stoichiometric composition, it was clear that the inner region of sample 2048HT had free carbon deposited between the coating layers, which resulted in stunted grain growth compared to the outer region (Fig. 2).

3.2 Irradiated Microstructure

Initial visual observations of irradiated samples during extraction and sorting at the IMET facility showed significant differences in the coating failure breakage between the 2 dpa at 800°C, 2 dpa at 1250°C and 6 dpa at 1250°C samples. Majority of 2 dpa samples had suffered from coating failure (breakage) post- irradiation. Statistically more samples irradiated at 2 dpa at 800°C had suffered failure compared to samples irradiated to- 2 dpa at 1250°C and 6 dpa at 1250°C. The reason for this coating failure in the 2dpa, 800°C samples was not resolved and could only be attributed to damage during handling of the 2 dpa samples prior to or after irradiation. Further instances of moderate to heavy grain dislodging in coatings was observed during sample preparation in the 2 dpa samples indicating damage. However it could not be clearly attributed to irradiation effects at this point in time. It is noted here that only samples with intact coatings were selected for further microstructural observations.

Observations of sample 3002 (Fig. 4 through Fig. 7) revealed no significant change from the preirradiation coating microstructure except in samples irradiated to 2 dpa at 800°C, which showed the formation of a layered structure not found in the pre-irradiation microstructure (Fig. 4). Figure 5 shows a higher magnification micrograph clearly revealing a layered microstructure in these samples. Annular microcracks were observed in these samples.

Micrographs of sample 3002HT indicated some grain coarsening attributed to irradiation (Fig. 4 through Fig. 7). The grain growth seemed to increase slightly with dpa as evidenced by the micrographs. However, similar to its predecessor, the 3002HT samples irradiated to 2 dpa at 800°C showed intergranular cracking and noticeable grain dislodging during preparation, indicating significant microstructural damage. From these observations, it was not possible to conclude if this was the effect of irradiation or of pre-irradiation damage during fabrication and handling.

Irradiated microstructure of sample 2048HT revealed a preserved pre-irradiation two region layered structure with additional grain growth (Fig. 4 through Fig. 7). However samples irradiated to 2 dpa at 1250°C showed significant cracking in the inner region which could not be clearly explained. In 2048HT samples, the outer layer had experienced a higher degree of grain growth compared to the inner layer resembling the 3002HT samples. The extent of grain growth in the inner region was significantly lower indicating the presence of interlayer carbon as previously hypothesized resulting in inhibition of large scale grain growth during irradiation. This hypothesis was again confirmed during subsequent microstructural observations of carbon rich 2003HT sample, which indicated no significant grain coarsening after irradiation. The clearly visible carbon interlayer had inhibited grain growth in these carbon rich ZrC samples, which could be attributed to increased separation leading to suppression of grain coalescence. All the 2003HT samples clearly retained the pre-irradiation layered structure.

Fig. 2. Microstructures of As-fabricated and Heat treated samples (Unirradiated)

Fig. 3. Microstructure of Sample 2048HT before Heat Treatment

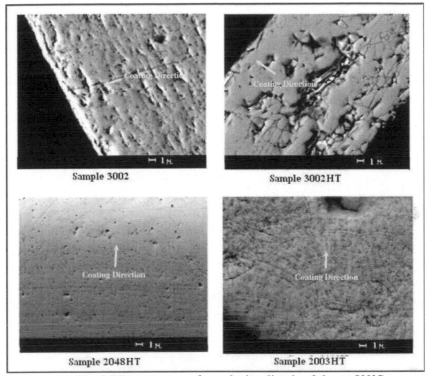

Fig. 4. Microstructure of samples irradiated to 2 dpa at 800°C

Fig. 5. Higher magnification micrograph of Sample 3002 irradiated to 2 dpa at 800°C

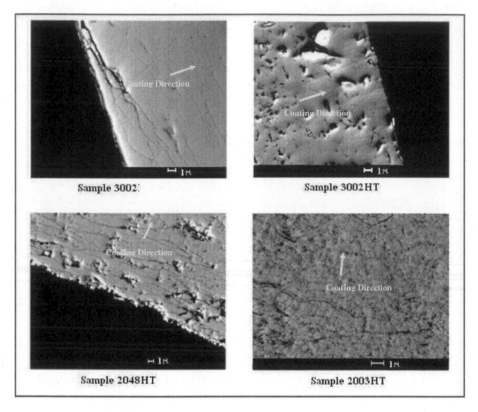

Fig. 6. Microstructure of samples irradiated to 2 dpa at 1250°C

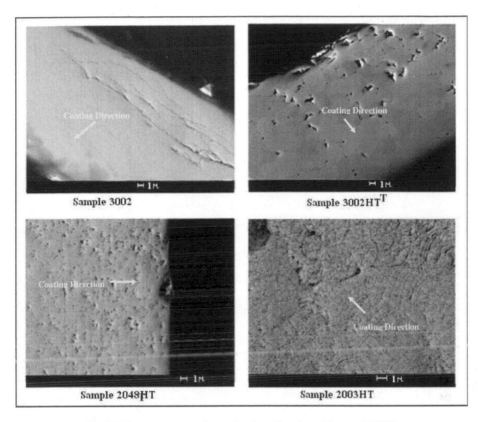

Fig. 7. Microstructure of samples irradiated to 6 dpa at 1250°C

4.0 Conclusions and Future Work

Microstructural comparison of as-fabricated, heat treated and irradiated CVD ZrC coated surrogate YSZ TRISO fuel was performed as a part of an International Nuclear Energy Research Initiative partnership between JAEA and ORNL. Surrogate fuel particles were irradiated in the HFIR facility at ORNL and subject to post-irradiation examination at the ORNL-LAMDA facility leading to the following observations:

As-fabricated and Heat Treated Microstructures

Nominally stoichiometric samples with as-fabricated, , non-layered microstructure experienced severe grain growth during heat treatment, whereas hyperstoichiometric samples with as-fabricated layered microstructure did not exhibit the same degree of grain size enhancement. This was attributed to the deposition of free carbon between the coating layers during fabrication in the hyperstoichiometric samples. Further, coatings with nominally stoichiometric ZrC (2048HT) had a distinctive microstructure with two regions each consisting of near isotropic grains of two sizes. The inner layer closer to the IPyC had finer grains than the outer layer, typical of carbon rich 2003HT samples. The grain structure in the outer region resembled the nominally stoichiometric 3002HT samples. This two region structure indicated changes in conditions in the middle of a particular coating run.

Irradiated Microstructures
1. As-fabricated samples without a layered structure had undergone further grain growth during irradiation. Microstructural damage was observed in these specimens indicating further possible degradation in mechanical properties.
2. Samples with a two region layered grain structure and nominally stoichiometric composition (2048HT) experienced a lesser degree of grain growth during irradiation similar to the effect of heat treatment. The effect of irradiation on the inner region was similar to carbon rich ZrC (2003HT) and the outer layer was similar to nominally stoichiometric ZrC (3002HT). The stunted grain growth in the inner layer was attributed to the presence of interlayer carbon.
3. The microstructures of samples with carbon rich ZrC were clearly the least affected by irradiation as with heat treatment. These samples underwent minimal grain growth compared to the other three samples, an effect which could be attributed to the excess carbon rich bands between the coating layers.

These preliminary microstructural observations indicated that, in general, the currently proposed ZrC coatings where severe grain growth was observed could be deemed mechanically unsuitable as a coating material under the described heat treatment and irradiation conditions.

Further studies including nanoindentation and crack propagation experiments planned under this program are expected to reveal changes in some mechanical properties of both irradiated stoichiometric and carbon rich samples. Additionally, detailed grain orientation studies are planned to obtain more information on the grain texture and to statistically gauge the extent of grain growth in the irradiated specimens.

5.0 Acknowledgements
This work was supported by the International Nuclear Energy Research Initiative (Project Number: 2006-001-J) and the Advanced Gas Reactor projects under the aegis of the US Department of Energy. The work described above is a collaborative effort of the Oak Ridge National Laboratory and the Japan Atomic Energy Agency. The authors wish to acknowledge the staff at ORNL-LAMDA and ORNL-IMET facilities for their assistance at various stages of the project. The authors also gratefully acknowledge the assistance rendered by the ORNL SHaRE user program in allowing the use of XL-30 and JEOL 6500F scanning electron microscopes for the microstructural observations.

6.0 References
1R.M.Versluis, F. Venneri, D.Petti, L.L.Snead and D.McEachern, "Project Deep-Burn: Development of Transuranic fuel for high temperature helium cooled reactors", *Proceedings of the 4th International Topical Meeting on High Temperature Reactor Technology*, Washington DC, 2008.

2 K. Minato, T. Ogawa, K. Fukuda, S. Kashimura, M. Shimizu and Y. Tayama, "Fission Product Palladium-Silicon Carbide Interaction in HTGR Fuel Particles", *J. Nucl. Mat.* 172, 184 1990.

3 T.Ogawa and K.Ikawa, "Reactions of Pd with SiC and ZrC", High Temp Sc., 22, 1986

4 T. Ogawa and K. Ikawa , "High-temperature heating experiments on unirradiated ZrC-coated fuel particles *J. Nucl. Mat.*, 99,1, 1981.

5 G.W. Keilholtz, R.E. Moore and Osborne M.F., "Fast neutron effects on the carbides of Titanium, Zirconium, Tantalum, Niobium and Tungsten", *Nuclear Applications*, 4, 1968.

6 D.A. Dyslin, R.E.Moore, and H.E.Robertson, "Irradiation damage to non-fissionable refractory materials", *ORNL Report -4480*, 1969.

7. K.H. Valentine, F.J. Homan, E.L. Long, T.N. Tiegs, B.H. Montgomery, and R.L. Beatty, "Irradiation Performance of HTGR Fuel Rods in HFIR Experiments HRB-7 and -8", *ORNL Report # 5228*, 1977.

8 Gan J., Yang Y., Clayton D., Todd A., "Proton irradiation study of GFR candidate ceramics", *J. Nucl Mat.*, 389, 2, 2009.

9 Y. Yang, A.D. Clayton, S. Hannah, M. Brandon, R.A. Todd, "Microstructure and mechanical properties of proton irradiated zirconium carbide", Journal of Nuclear Materials, 378, 3, 2008.

10 L.L. Snead, Y. Katoh, and S.Kondo. "Effects of fast neutron irradiation on zirconium carbide", *J. Nucl Mat.*, To Be Published, 2009.

11 D. Gosset, M. Dollé, D. Simeone, G. Baldinozzi, and L. Thomé., "Structural evolution of zirconium carbide under ion irradiation", *J.Nucl Mat.*, 373, 1-3, 2008.

12 J. Aihara, S.Ueta, A.Yasuda, H. Ishibashi, T. Takayama, K.Sawa, and Y. Motohasi, "TESM/STEM observation of ZrC-coating layer for Advanced High Temperature Gas cooled reactor fuel", *J.Am.Ceram.Soc*, 90, 12, 2007.

13. J. Aihara, S.Ueta, A Yasuda, H. Ishibashi, T. Takayama, K.Sawa, and Y. Motohasi, "TESM/STEM observation of ZrC-coating layer for Advanced High Temperature Gas cooled reactor fuel , Part II", *J.Am.Ceram.Soc*, 92, 1, 2009.

14. J. Aihara, S.Ueta, A.Yasuda, H. Takeuchi, Y.Mosumi, K.Sawa, and Y. Motohasi, "Effect of heat treatment on TEM microstructrues of Zirconium Carbide coating layer in fuel particle for advanced high temperature gas cooled reactor", *Mat.Transc*, 50, 11, 2009.

Advanced Coating
Processing

INTRODUCTION TO HIGH VELOCITY SUSPENSION FLAME SPRAYING (HVSFS)

R. Gadow, A. Killinger, J. Rauch and A. Manzat
Institute for Manufacturing Technologies of Ceramic Components and Composites
University of Stuttgart
Stuttgart, Germany

ABSTRACT

High Velocity Suspension Flame Spraying (HVSFS) has been developed for thermal spraying of suspensions containing micron, submicron and especially nanoparticles with hypersonic speed. For this purpose, the suspension is introduced directly into the combustion chamber of a modified HVOF torch thereby featuring a safe work place conditions while handling nano powders-

The aim in mind is to achieve dense and well adherent coatings with a refined microstructure. Especially from nanostructured coatings superior physical properties are expected for many potential applications. Direct spraying of suspensions offers flexibility in combining and processing different materials. It is a cost saving process and allows the allocation of entirely new application fields.

The paper gives an overview of the HVSFS method and will present some actual results that have been achieved by spraying nanooxide ceramic materials like Al_2O_3, TiO_2, 3YSZ and Cr_2O_3.

INTRODUCTION

Thermal spray technology has been applied for decades with great success and still bears a great potential for new applications in many industrial fields. One of the great benefits is the broad variety of spray materials that can be applied to all kind of substrate materials.

However, standard spray techniques face restrictions regarding their coating thickness and resulting microstructure. From the fact that the processed spray materials (rods, wires or standard spray powders) result in spray particles with a certain particle size that cannot be reduced, there are limitations to decrease coating thickness significantly below a value in the range of 30 µm because individual splats are too coarse to form a homogeneous and sealed coating structure. For most applications, a coating thickness of 200 to 300 µm matches well with the technical requirements, however, there are new applications, where thin coatings in the range of 5 to 50 µm would be of great interest, e. g. in the field of sensor technology, development of solid electrolytes (e. g. for fuel cells) and for a new generation of nanostructured tribologically functional coatings.

A reduction of particle sizes below 5 µm means an extraordinary effort in powder feeding technique, as fluidization of the spray powder gets more and more challenging with a decreasing particle size. Therefore suspensions seem to offer a promising solution of this problem. Moreover, using a suspension opens up a completely new field of specially designed spray materials, including nanopowders that are currently not available on the spray powder market. The processing of a nanopowder by means of the standard thermal spray procedure, first of all requires the agglomeration of the nanopowders by means of a spray drying process in order to form spray particles with appropriate grain size that are suitable for a standard powder feeding device. These powder qualities are yet expensive and not available on the market for every combination of materials. Oxide based materials are supplied by several companies, whereas cermet based powder materials containing nanosized hardphase particles like carbides or borides are relatively rare and expensive. Spraying these powder types, however, leads to spray particles with particle diameters comparable to standard spray powders. The aim to form thin film spray coatings with a refined internal coating structure cannot be solved by this approach, although (in the case of mixed phase powders) a dispersion of submicron or even nanoscaled hardphases can be realized. A general problem is the fact that during the melting process in the flame the agglomerated particles lose their nanostructure (in case of a single phased agglomerate) so it is not yet clear if there is any benefit compared to using standard powders.

Suspension spraying has been carried out using plasma spraying, flame spraying and HVOF[1234]. In the case of plasma spraying, the injection of the liquid generally is performed perpendicular to the plasma flame. In the case of conventional flame and HVOF spraying, an axial injection into the flame or combustion chamber respectively is preferred as it circumvents the problem of strong flame disturbance causing instabilities in the spray process[1].

Introducing a liquid into the spray process raises some basic problems. Any liquid, being water or an organic solvent, starts to evaporate rapidly when introduced into the flame causing two important effects. a) A significant cooling of the flame and b) due to the evaporation process the expanding vapor causes strong disturbance of the free expanding hot gas stream, especially in plasma spraying. When introduced into a combustion chamber a significant rise of the combustion chamber pressure will occur as can be observed in the HVSFS process.

Particle size strongly influences the particle thermal history. Small particles rapidly heat up but they also rapidly cool down resulting in a fair different behavior concerning the splat formation on the surface. It also affects interlamellar adhesion of the splats and thereby influences the mechanical properties (e. g. Young's modulus, hardness) of the coating. When thermally spraying suspensions, the liquid is fragmented through an atomization process into small droplets not necessarily having the size of the primary particles (e. g. nanoparticles), this issue has been extensively discussed in literature[5]. It has been reported for plasma spraying of suspensions[6] but can also be observed in the HVSFS process, as shown by the authors in[4]. The goal is to optimize the atomization in the combustion chamber to achieve suspension droplets as small as possible. This leads to finer splats in the coating with the improvements described above.

THE HVSFS PROCESS

Injecting a liquid feedstock into the flame or combustion chamber respectively implies a different combustion process compared to the conventional HVOF process. If suspensions are used as carrier medium, an optimization of the injection system and the combustion chamber of the HVSFS torch are required.

Regarding the injection (of the ceramic suspension) it has to be considered that the liquid feedstock must overcome the pressure in the combustion chamber. Coating experiments showed that a standard injection nozzle tends to clog due to the unfavorable flow speed distribution. To optimize the suspension injection, different injection nozzle designs have been investigated and tested. The simulated cross-section speed profiles of the flow in a cylindrical and a conical shaped suspension injection nozzle are depicted in Figure 1.

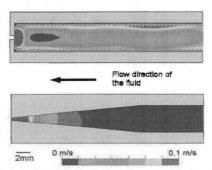

Figure 1. Cylindrical design (top) vs. improved design (bottom)

In the cylindrically shaped nozzle areas of reduced flow speeds can be observed behind the outlet to the combustion chamber, which can cause clogging of the outlet. The conically shaped nozzle prevents clogging of the suspension before the outlet into the combustion chamber because it provides for a uniform flow speed throughout the nozzle. The flow speed steadily increases towards to the combustion chamber reaching the maximum at the outlet. The suspension outlet diameter into the combustion chamber of both mentioned nozzles is 0.3 mm.

A further approach to enhance the HVSFS process is optimizing the design of the combustion chamber. Figure 2 displays CFD-simulations of the gas flow of a cylindrical and a conical design using propane as fuel gas.

The implementation of a cylindrical shape of the combustion chamber is widely spread for the conventional HVOF technique. The gas flow simulation of the cylindrical shape, top of Figure 2, shows turbulent gas combustion. To achieve a fast and complete combustion of the fuel gas, a turbulent gas flow is helpful. However, a turbulent gas flow during the HVSFS process can impair the coating process due to the fact that the nanoparticles in the suspension follow the streamlines of any turbulence more easily compared to conventional spray powder particles. This often leads to an uncontrolled deposition of particles in the combustion chamber and expansion nozzle respectively. To overcome this problem, new nozzle and combustion chamber designs have been developed. The idea in mind is to reduce turbulence inside the combustion chamber and keep the suspension droplets in the center of the gas jet.

In contrast to the cylindrical design, the conical shape shows laminar gas combustion of propane (Figure 2). This is convenient for processing submicron and nanoparticle-containing suspensions. On the other hand, due to the missing turbulence, an incomplete combustion of the propane can be observed. As a result, the heat release of the exothermal reaction is decreased.

Fuelgas inlets

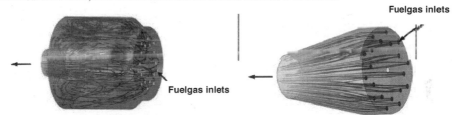

Fuelgas inlets

Figure 2. CFD-simulation of gas flow lines for two different combustion chamber designs.

The flame temperature is also influenced by the suspension. Any liquid, water as well as organic solvents like isopropanol, fed into the flame of the combustion chamber evaporates instantly. In consequence of the evaporation process the expanding vapor causes a rise of the combustion chamber pressure and a considerable change of the flame temperature. Using water as fluid of the suspension causes a reduction of the flame temperature, because heat is consumed by the evaporating water (enthalpy of vaporization). A downside of this cooler flame is the reduced capability to melt materials with a high melting point. Using isopropanol as a solvent for the suspension in the HVSFS process noticeably increases of the flame temperature due to the released energy (enthalpy of combustion).

Table I. Comparison of density, evaporation temperature and evaporation and reaction enthalpies of the solvents used to fabricate the suspensions [7 8]

	Water	Ethanol	Isopropanol
Density, ρ_1 [g/cm^3] at 20°C	0.9982	0.7893	0.7855
Evaporation temperature [°C]	100	78.3	83
Evaporation enthalpy [kJ/g]	2.26	0.84	0.67
Evaporation enthalpy [kJ/mol]	40.72	38.57	40.17
Reaction enthalpy [kJ/g]	-	29.7	33.39
Reaction enthalpy [kJ/mol]	-	1366.2	2003.7

Processing an isopropanol based suspension leads to an increase of the thermal energy in the HVSFS system. This allows for a compensation of the decreased heat release of the exothermal reaction by using a conical shaped combustion chamber. The additional enthalpy and the direct combustion of the isopropanol on the surface of the nanoparticles also improves the heat transfer into the particle which than leads to a faster melting of the submicron or nanoscale materials.

This means, that even with an incomplete combustion of the fuel gas, nano materials with a high melting point, like n-Cr$_2$O$_3$, can be melted. In addition, the improved trajectory of the submicron and nano particles prevents the uncontrolled deposition of particles in the combustion chamber and expansions nozzle. In Figure 3 the cylindrical and the conical combustion chamber design with the attached expansion nozzle is pictured.

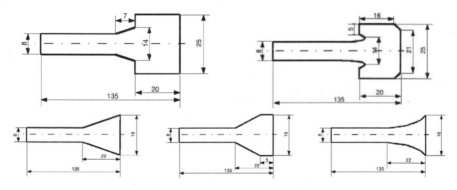

Figure 3. standard combustion chamber designs (top) vs. new conical combustion chamber designs (bottom)

Common process parameters of HVSFS

A main difference between standard HVOF spray process and HVSFS is the significantly shortened stand-off distance. Typical spray distances are in the range of 100 to 150 mm.

All suspensions are sprayed using propane or ethene as a fuel gas. When choosing propane as a fuel gas, it was found that an oxygen-fuel ratio adjusted to approx. 5.4 achieved best deposition rates. In case of ethene the optimum oxygen-fuel ratio was approx. 3.3. It should be kept in mind that flame temperature is strongly influenced by the presence of isopropanol in the flame. It also heavily influences the oxidation behavior of the flame.

Basically, using acetylene can be of advantage when spraying oxide materials, as it delivers higher flame temperatures[9]. However, due to the insertion of liquid fuel, the combustion chamber pressure rises significantly, exceeding the maximum gas line pressure of 1.5 bars that is allowed for acetylene, therefore expensive high pressure lines would be necessary to run a stable HVSFS process. Ethane is a reasonable alternative as it shows a flame temperature of 2924 °C as opposed to propane having 2828 °C[10].

EXPERIMENTAL

The typical configuration of the HVSFS equipment is shown in Figure 4. It consists of a suspension container connected to a pump feeder system. A flexible suspension line feeds the suspension axially and concentrically into the combustion chamber of a modified HVOF gun. A second container holds the pure solvent to flush the suspension line and injection nozzle when shutting down the process. The HVSFS torch is based on a TopGun-G system (manufactured by GTV) and has been modified to meet suspension spray requirements as described in the chapter above.

For all suspensions, isopropanol was chosen as the organic solvent. Although stabilization of particles is more difficult than in aqueous media, this alcohol delivers fairly high enthalpy values during the combustion process. In case of the TiO$_2$, the suspension also contained water (10 wt.-% of the liquid phase) to optimize the viscosity by using the dispersing agent DOLAPIX CE64.

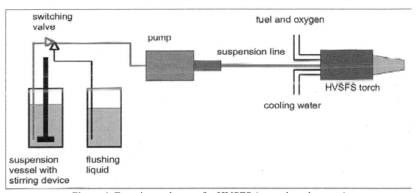

Figure 4. Experimental setup for HVSFS (pump based system)

PREPARATION OF THE SUSPENSION

Table I summarizes all powders used in this work. In a first step the powders were attrition milled in isopropanol solvent to force deagglomeration and particle breakup respectively. This is necessary because most nanopowders show agglomerates with grain sizes which might clog the suspension injection nozzle (diameter 0.3 to 0.5 mm). To remove milling balls and possible agglomerates, the suspensions were sieved after milling process with a mesh of 50 μm. The suspensions were then applicable for spraying. The solid content of all suspensions in this work was in the range of 15 to 20 wt.-%. A summary of the suspension compositions is given in Table II.

Table II. Summary of applied powder materials

Brand name (manufacturer)	Chemical composition	Primary grain size [nm]	Phases analysis derived from XRD
TM-DAR (Taimei Chemicals)	n-Al_2O_3	150	γ-Al_2O_3
Aeroxide P25 (Degussa)	n-TiO_2	15	anatase, rutile
Nanofiller (DGTec)	n-Cr_2O_3	100	eskolaite
VP PH (Degussa)	n-3YSZ	15	tetragonal, monoclinic

Table III. Composition of the different suspensions

Suspension	Solid content [wt.%]	Isopropanol content [wt.%]	Additives
n-Al_2O_3	20	80	HNO_3
n-TiO_2	10	90 / 10	DOLAPIX CE64
n-Cr_2O_3	15	85	HNO_3
n-3YSZ	20	80	HNO_3

COATING EXPERIMENTS

All suspensions were sprayed using propane or ethene as fuel gases, relevant spray parameters are summarized in
Table III. When choosing propane as a fuel gas, it was found, that an oxygen/fuel ratio adjusted to approx. 5.4 achieved best deposition rates. In case of ethene the oxygen/fuel ratio was approx. 3.3. It should be kept in mind, that flame temperature is strongly influenced by the presence of isopropanol in the flame. It also strongly changes the oxidation behavior of the flame.

Coatings were applied either on planar mild steel (St37) or aluminum alloy (AlMg3) samples (50 x 50 mm) and were mounted vertically on a sample holder. The HVSFS torch was operated on a six axis robot system using a simple meander kinematic with 1 mm offset. Sample cooling was performed using two air-nozzles mounted on the torch. Substrate preparation prior to spraying included degreasing the surface with acetone. For all substrates a mild injection grit blasting operation was carried out using white corundum particles (FEPA 120; 100 – 125 µm) and 5 bar compressed air. Spray parameters are given in Table IV.

Table IV. Parameters for HVSFS (P= propane, E= ethene)

Suspension	Fuel gas [slpm]	O_2 [slpm]	Torch speed [mm/s]	No. of cycles	Spray dist. [mm]	feeding rate [g/min]
n-Al_2O_3	P 65	350	600	10	150	10
n-TiO_2	P 50	220	800	10	100	8
n-Cr_2O_3	P 65	350	600	5	125	8
n-Cr_2O_3	E 90	300	600	5	125	8
n-3YSZ	E 90	300	600	10	120	10

CHARACTERIZATION OF THE HVSF SPRAYED COATINGS

From all samples, cross sections were prepared for image analysis using light microscopy and/or SEM. Surface topography was taken in SEM to reveal the structure of single splats. Microhardness was measured on coating cross section using a Fisherscope equipped with a Vickers indenter. All measurements are performed dynamically (universal hardness) according to DIN 50359 1. The given values are the average of 10 indentations. Coating porosity was determined from digital image analysis (Software AQUINTO). Surface roughness was measured using a contact profilometer (Perthometer Concept, Mahr), R_z and R_a values were determined according to DIN EN ISO 4287. All results are summarized in Table V.

Al$_2$O$_3$ coatings

Aluminum oxide suspension was HVSF sprayed on aluminum substrate using propane as fuel gas. A conical shaped injection nozzle was used. The feeding rate was approx. 10 g/min and the coating thickness was about 100 μm. Deposition rate was approx. 10 μm per transition. The alumina coating exhibits a good adhesion to the aluminum substrate, Figure 5a. The porosity was approx. 5 %, micro hardness was in the range of 600 to 900 $HV_{0.1}$.

The coating surface shows fully molten droplets in the range of 1 to 5 μm, in the intersplat region small spherical particles with a size of 200 up to 500 nm are visible, Figure 5b. Surface roughness is $R_z = 3.14$ μm and $R_a = 0.58$ μm.

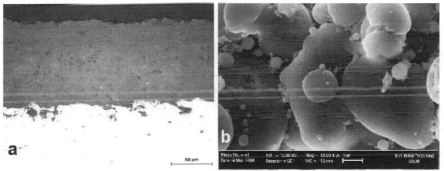

Figure 5. a) Optical micrograph of HVSF sprayed Al$_2$O$_3$ coating. b) SEM image of the coating surface.

TiO$_2$ coatings

The titanium oxide suspension was HVSF sprayed on aluminum substrate using propane as a fuel gas. The feeding rate was approx. 8 g/min., a conically shaped 0.3 mm nozzle was used. The applied coating thickness is approx. 30 μm, deposition rate was approx. 4 μm/cycle. In cross section, the TiO$_2$ coating appears with a very low porosity, Figure 6a. The coating appears homogeneous, no internal structure is visible in cross section and no unmolten particles are detected. The SEM surface image reveals fully molten splats ranging from a few 100 nm up to 10 μm with a good inter splat connection, Figure 6

From the XRD data it can be stated that the main fraction of the coating consists of anatase (approx. 75 %), the rest of the material consists of rutile.

Figure 6 a) Optical micrograph of HVSF sprayed TiO_2 coating. b) SEM image of the coating surface.

Cr₂O₃ coatings

Chromium oxide suspension was HVSF sprayed with propane and ethene to compare the influence of the fuel gases. Suspension feeding rate, torch set up, robot kinematics and substrate were identically for both coatings, refer to Table IV. The n-Cr_2O_3 suspension was sprayed on steel substrates with a feeding rate of approx. 8 g/min using a cylindrically shaped 0.4 mm nozzle. Deposition rate was in the range of 2 μm/cycle for both fuel gases. The achieved coating has a thickness of 10 to 15 μm, exhibits good adhesion to the steel substrate and shows a porosity of approx. 5 % in case of propane and approx. 10 % in case of ethene, refer to Figure 7a and Figure 7b, for porosity data see Table V.

The microhardness of the propane sprayed coating was in the range $1400 - 1800 \, HV_{0.01}$, ethene sprayed in the range of $1100 - 2000 \, HV_{0.01}$. In case of the ethene sprayed samples, the lower value of the measured microhardness is caused by the higher porosity in the coating. There is a fairly high variance of the measured values due to the low indenter forces.

In SEM, coatings show different surface structures. The surface of the coating sprayed with propane appears coarser with fully molten droplets in the range of 1 to 3 μm, Figure 8a. A refined splat structure is visible on the ethene sprayed coating surface, Figure 8b. Small spherical droplets in the range of 100 nm are present together with larger droplets (1 to 3 μm). The surface roughness of both coatings is comparable; Rz = 2.75 μm, Ra = 0.47 μm for the propane sprayed and Rz = 2.79 μm, Ra = 0.48 μm for the ethene sprayed.

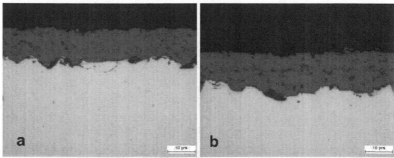

Figure 7. Optical micrograph of HVSF sprayed chromium oxide cross sections. a) Sprayed with fuel gas propane. b) Sprayed with fuel gas ethane

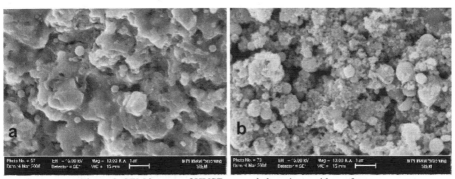

Figure 8. SEM images of HVSF sprayed chromium oxide surfaces.
a) Fuel gas propane. b) Fuel gas ethane

3YSZ coatings

The YSZ suspension was flame sprayed on steel substrate with a feeding rate of approx. 10 g/min using a 0.4 mm conical nozzle. In this case, ethene was chosen as the fuel gas to achieve a high flame temperature. The deposition rate was in the range of 3 μm/cycle. In cross sections the YSZ coating appears dense and homogeneous. The thickness of the coating is about 30 μm Figure 9a. In SEM images stretched pores with a diameter of about 200 nm are visible, Figure 9b. The microhardness of the coating was measured 766 $HV_{0.03}$. From XRD it can be stated that the coating consists of a pure tetragonal phase.

The surface of the 3YSZ coating appears as a fully molten splat structure with splats in the range of 1 to 3 μm and a well established intersplat connection. Fine spherical particles with a diameter of 50 to 200 nm are distributed between the larger agglomerates, Figure 10a and Figure 10b. The surface roughness was measured to R_z = 8.68 μm and R_a = 1.75 μm.

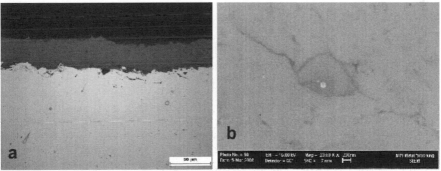

Figure 9. a) Light microscope image of a HVSF sprayed 3YSZ coating;
b) SEM image of HVSF sprayed 3YSZ, pore structures are in the range of 50 nm

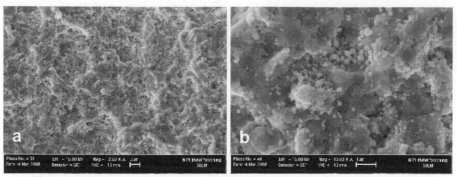

Figure 10. SEM images of the surface of HVSF sprayed 3YSZ at different magnifications

HAp coatings

The hydroxyapatite (HAp, $Ca_5(PO_4)_3OH$) has been sprayed using water- and diethylene glycol-based suspensions using a conically shaped injection nozzle together with a conical expansion nozzle. The deposition rate reached up to 15 μm/cycle, leading to a quick coating build-up, even when spraying up to 2 mm thick coatings for some specific applications.

The Hap retained its crystalline structure throughout the coating process as can be seen from the XRD analysis in Figure 11.

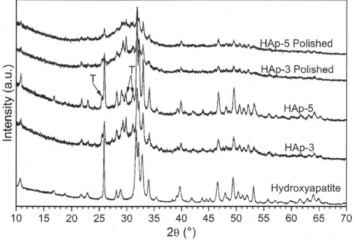

Figure 11. XRD patterns of the polished and unpolished HAp coatings compared to pure hydroxyapatite. (Label T = tetracalciumphosphate, all other peaks = hydroxyapatite)[11]

Table V. Mechanical and structural data of the HVSF sprayed coatings

Spray material	Vickers hardness	Coating porosity (%)	R_z [μm]	R_a [μm]	Phase composition derived from XRD
n-Al$_2$O$_3$	620 – 880HV$_{0.1}$	5	3.14	0.58	mainly γ
n-TiO$_2$	approx. 1000 HV$_{0.05}$	0 - 0.5	3.5	0.65	mainly anatase
n-Cr$_2$O$_3$ propane	1400 – 1800 HV$_{0.01}$	4 – 5	2.75	0.47	hexagonal
n-Cr$_2$O$_3$ ethene	1100 – 2000 HV$_{0.01}$	7 - 10	2.79	0.48	hexagonal
n-3YSZ	715 – 816 HV$_{0.03}$	< 1	8.68	1.75	tetragonal

APPLICATIONS

The novel HVSFS process has already been deployed for the deposition of high-performance coatings for tribological and other industrial applications, where comparatively thin and smooth coatings are beneficial.

Cutter Blades

One example for the advantages of HVSF sprayed coatings are cutter blades for hedge trimmers. As a result of the smooth as-sprayed surface (R_a = 2.5 μm, R_z = 8 μm), the Cr$_2$O$_3$-coatings can be deployed directly without the need for any mechanical post-treatment. In addition, the chromia coatings are very resistant against corrosion and wear, the latter especially due to the high coating hardness (2000 . The process time for applying a 30 μm thick coating to four cutter blades is about 2 minutes with potential for some further optimization and therefore represents a fairly quick process for refining the cutter blades.

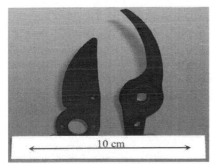 

Figure 12. Cr$_2$O$_3$ HVSFS coated cutter blades (left) and cross section micrograph (right)

Solid Oxide Fuel Cell

HVSF sprayed 8YSZ coatings have been tested for a possible application as thin electrolyte layer in SOFCs as well as material for the electrodes. The HVSFS process allowed for the deposition of dense and gas-tight coatings having a thickness of only 20 μm. These thin layers therefore contribute to the development of SOFCs with a reduced ionic resistance as well as a reduced gas leakage.

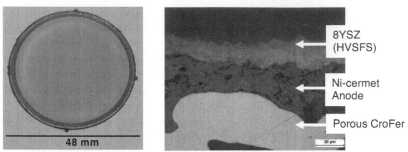

Figure 13. SOFC test sample (left) and cross section micrograph (right)

Cylinder Liner and Engines

HVSFS coatings have also been applied to cylinder liner running surfaces of internal combustion engines[12]. For this application, a HVSF sprayed TiO_2/TiC- coating has been chosen because it features dry-lubricating phases that aid in reducing the friction coefficient in deficient lubrication states of the engine run. In addition the comparatively high hardness (800 $HV_{0.1}$) also contributes to a wear reduction.

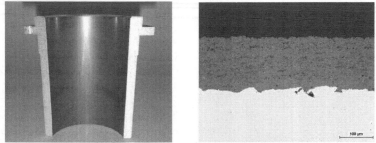

Fig. 14. TiO_2/TiC-coated cylinder liner (left) and cross section micrograph of the coating (right)

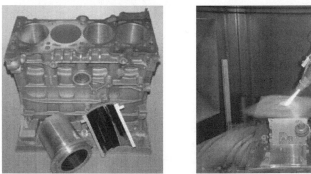

Figure 15. Inline-4-cylinder engine and cylinder liners (left) and rotating inline-4-cylinder engine during the coating process (right)

The coating mainly contains nanoscale TiO_2 (primary particle size = 20 nm), the fine coating microstructure and the resultant micro porosity on the surface also support a hydrodynamic friction state during operation of the engine.

Biocompatible Coatings

The HVSFS technique can be very useful for the production of high-strength biocompatible coatings on metallic orthopedic implants. These coatings generally consist of hydroxyapatite (HAp, $Ca5(PO4)3OH$), which is chemically similar to the mineral component of human bones and hard tissues[13]. It is indeed able to support bone in-growth and osteointegration when used in orthopedic, dental and maxillofacial applications[14][15][16]. Moreover, HAp coatings have the capacity of cutting down the healing process of implants with a metallic surface. In Figure 16 a SEM micrograph of the polished cross-sections of a HAp HVSFS sprayed coating is displayed.

The microstructure of the coating depends primarily on the liquid phase of the suspension: water-based ones yield more porous coatings, retaining some more unmelted crystalline phase than the diethylene glycol-based ones. In all cases, the surface of the coatings shows higher concentration of hydroxyapatite than their interior, because of the deposition of overspray particles containing partially unmelted agglomerates.

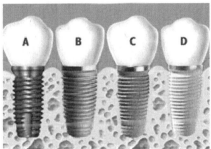 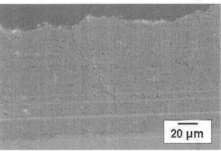

Figure 16. left: Machined Titanium (A), Acid Etching Titanium (B), Titanium Plasma Spray Coating (C), HAp Coated Titanium (D); right: SEM micrograph of the polished cross-sections of a HAp HVSFS sprayed coating

CONCLUSION

All HVSF sprayed coatings exhibit a refined splat or micrograin structure that is at least one order of magnitude below standard spray coatings. This is also valid for the pore structure. Due to this fact all HVSF sprayed coatings exhibit remarkably low values of surface roughness. From all investigated coatings, the titanium and zirconium oxide coatings exhibit lowest porosity values.

Generally, there is a fairly high variance of the measured hardness values due to the reduced coating thickness in case of chromium and zirconium oxide. Further investigations will include nanoindenter analysis to achieve more precise values for coatings below 50 µm.

For all coatings, there is no visible evidence that particles were insufficiently molten. So far, there is no clear tendency that porosity would directly correspond to the melting point of the sprayed material, titanium oxide (melting point 1800 °C) and 3YSZ (melting point 2700 °C) show the highest density values. Deposition efficiency is in the range of 3 to 4 µm/cycle for all investigated suspensions. So far, using ethene as a fuel gas providing a higher flame temperature did not show any advantage concerning the deposition efficiency and coating porosity achieved in the HVSFS process.

From XRD data, it is found that a crystallite structure in the coating significantly exceeds the size of the initial nanopowders. However, from our earlier publication it was found, that spraying of mixed nanooxides can preserve a nanostructure in the coating[11][12].

HVSFS coatings can have beneficial effects compared to conventional thermal spray coatings in common applications like functional coatings for tribological[17] or bio-medical[18] applications. In addition, new application areas can be opened in consequence of the lower coating thicknesses, smoother surfaces and improved coating properties achieved by the novel concept of directly spraying nanoscale materials by means of HVSFS[19][20][21].

SUMMARY

The HVSFS method was evaluated to manufacture coatings from n-Al_2O_3; n-TiO_2, n-Cr_2O_3, n-3YSZ nanooxides and HAp. Appropriate suspensions were prepared in-house using isopropanol, water and diethylene glycol as a solvent. All materials were sprayed using propane or ethene as a fuel gas.

For all investigated materials, homogeneous coatings featuring a significantly lower surface roughness and a highly refined microstructure compared to standard spray coatings could be manufactured. So far measured micro hardness values are comparable or slightly lower than those of standard coatings, but further measurements are required to characterize the coatings more precisely. All coatings clearly show a crystalline structure in XRD.

The HVSFS coatings show a great potential for highly specialized future applications with improved qualities compared to standard HVOF coatings.

REFERENCES

[1] R. Gadow, A. Killinger, M. Kuhn and D. Lopéz: *Published patent application DE 10 2005 038 453 A1*, (2007)

[2] C. Delbos, J. Fazilleau, J.F. Coudert and P. Fauchais: Plasma Spray Elaboration of Finely Structures YSZ Thin Coating by Liquid Suspension Injection; Thermal Spray 2003: Advancing the Science & Applying the Technology, (Ed.) C. Moreau and B. Marple, *ASM International*, Materials Park, Ohio, USA, (2003), p 661-669

[3] T. Poirer, A. Vardelle, M. F. Elchinger, M. Vardelle, A. Grimaud and H. Vesteghem, Deposition of Nanoparticle Suspensions by Aerosol Flame Spraying: Model of the Spray and Impact Processes, Thermal Spray 2003: Advancing the Science and Applying the Technology, B.R. Marple and C. Moreau, Ed., May 5-8, 2003 (Orlando, FL), *ASM International*, (2003), Vol. 2,

[4] A. Killinger, M. Kuhn and R. Gadow, High-Velocity Suspension Flame Spraying (HVSFS), a new approach for spraying nanoparticles with hypersonic speed; *Surface and Coatings Technology* 201, (2006),

[5] E. Bouyer, F. Gitzhofer and M. I. Boulos, Powder Processing by Suspension Plasma Spraying; *Thermal Spray: A United Forum for Scientific and Technological Advances*, C.C. Berndt, Ed., Sept 15-18, 1997 (Indianapolis, IN), ASM International, (1998), 353 p

[6] R. Rampon, G. Bertrand, F.-L Toma and C. Coddet, Liquid Plasma Sprayed Coatings of Yttria Stabilized Zirconia for SOFC Electrolytes, Proceedings of the 2006 *International Thermal Spray Conference, ASM International*, Seattle, Washington, USA, (2006)

[7] C. Synowietz, K. Schäfer: *Chemiker-Kalender*, 3. Edition, Synowietz C., Schäfer K. (ed.), Springer-Verlag, 1984, ISBN 3-540-12652-x

[8] L. Pawlowski: Finely grained nanometric and submicrometric coatings by thermal sparaying: A review; *Surface and Coatings Technology*, 2008, 202, p 4318-4328

[9] R. Rampon, G. Bertrand, F.-L Toma and C. Coddet: Liquid Plasma Sprayed Coatings of Yttria Stabilized Zirconia for SOFC Electrolytes; *Proceedings of the 2006 International Thermal Spray Conference*, 2006, ASM International, Seattle, Washington, USA

[10]H. Kreye, F. Gärtner: High Velocity Oxy-Fuel Flame Spraying – State of the Art, new Developments and Alternatives; *Conference Proceedings 6th Colloquium 2003*, Erding, Germany

[11]G. Bolelli, V. Cannillo, R. Gadow, A. Killinger, L. Lusvarghi and N. Stiegler; Nanoceramic based High-Velocity Suspension Flame Spraying (HVSFS) of Hydroxyapatite coatings for biomedical applications; *12th Cells and Tissues International Congress on Bioceramics*, Faenza, (2009)

[12]R. Gadow, A. Killinger, A. Manzat; Innovative coatings for cylinder liner surfaces in high performance engines, *Motorsport Expotech 2nd editon*, Modena, Italy (2009)

[13]E. Wintermantel, H. Suk-Woo; Medizintechnik mit biokompatiblen Werkstoffen und Verfahren, *3. überarb. und erw. Aufl. Springer-Verlag*, (2002)

[14]P. Ducheyne, L. Hench, A. Kagan, M. Martens, A. Bursens, J. Mulier; Effect of hydroxyapatite impregnation on skeletal bonding of porous coated implants, *J Biomed Mater Res* (1980)

[15]K. de Groot, R. Geesink, C. Klein, P. Serekian; Plasma sprayed coatings of hydroxylapatite. *J Biomed Mater Res* (1987)

[16]C. Yang et al.; In vitro and in vivo mechanical evaluations of plasma-sprayed hydroxyapatite coatings on titanium implants: The effect of coating characteristics. *J Biomed Mater Res* (1997)

[17]H. Kreye, Vergleich der HVOF-Systeme - Werkstoffverhalten und Schichteigenschaften, *4. Kolloquium Hochgeschwindigkeitsflammspritzen*, (1997), p 13-21

[18]R. Gadow, F. Kern, A. Killinger; Manufacturing technologies for nanocomposite ceramic structural materials and coatings, *Materials Science & Engineering B*, in print

[19]R. Gadow, J. Rauch, N. Stiegler; Bioactive Nanoceramic coatings by supersonic suspension flame spraying (HVSFS); *1st Sino-German Symposium on Advanced Biomedical Nanostructures (SGSABN)*; Jena (2009)

[20]R. Gadow; A. Killinger; Verbesserung der Bauteileigenschaften durch multifunktionale Nanobeschichtungen, *Nanotechnologie; Seminar der Heilbronner technisch-wissenschaftlichen Vereine;* Hochschule Heilbronn (2008)

[21]R. Gadow, A. Killinger; HVSFS and HVOF suspension spraying; a review of the experimental achievements, *Les Rencontres Internationales sur la Projection Thermique, 4th RIPT and Suspension and Solution Thermal Spraying, 3th S2TS*, Ecole Nationale Supérieure de Chimie de Lille (ENSCL) Maison d'Activités Culturelles et de Colloques (MACC), University of Science and Technology of Lille (USTL), Lille (2009)

[22]V. Cannillo, G. Bolelli, R. Gadow, A. Killinger, L. Lusvarghi, A. Sola, N. Stiegler; High-Velocity Suspension Flame Sprayed (HVSFS) bioactive glass coatings; *Les Rencontres Internationales sur la Projection Thermique, 4th RIPT and Suspension and Solution Thermal Spraying, 3th S2TS*, Ecole Nationale Supérieure de Chimie de Lille (ENSCL), Maison d'Activités Culturelles et de Colloques (MACC), University of Science and Technology of Lille (USTL), Lille (2009)

ELECTROPHORETIC DEPOSITION OF OXIDE POWDER BY USING NON-FLAMMABLE ORGANIC SOLVENT

Hideyuki Negishi[1], Ai Miyamoto[2], Akira Endo[1], Keiji Sakaki[1], Hiroshi Yanagishita[1] and Kunihiro Watanabe[2]
1) National Institute of Advanced Industrial Science and Technology (AIST), AIST Central 5-2, 1-1-1 Higashi, Tsukuba, 305-8565, Japan
2) Tokyo University of Science, 2641 Yamazaki, Noda, 278-8510, Japan

ABSTRACT

The electrophoretic deposition (EPD) process is one of the ceramic powder assembling technologies. This technique has some advantages compared with other fabrication techniques in terms of the fabrication costs and the structural flexibility of the substrates. A solvent that has high relative dielectric constant is usually used as an EPD bath. In case of using organic solvents, high voltage is applicable to the EPD bath and the particles can migrate under the strong driving force. However, most of organic solvents have difficulties in practical use because of their flammability. In this study, we investigated the technological feasibility in preparation of various ceramic coatings by the applied high voltage EPD in a non-flammable organic solvent such as hydrofluoroether (HFE). We found that nonflammable HFE was useful for the EPD bath for the preparation of various oxide coatings. Ceramic powders, such as yttria-stabilized zirconia (YSZ), zeolite A4, silicalite, mesoporous silica (MPS) and silica powder could be deposited on the metal substrate under the applied voltage at 100V. In addition, the use of HFE can reduce the electric power consumption to 1/10000 or less compared to the use of organic solvents such as alcohol and acetone.

INTRODUCTION

In recent years, new technology to create additive properties by systematic arrangement of size-controlled fine particles in two or three dimensions has attracted great attention as a coating technology[1,2,3]. The wet-process fabrication is especially attractive because of its low cost and mass productivity. As for the systematic arrangement of particles on a substrate, controlling the interaction of the particle-particle and/or particle-substrate in the colloidal solution based on the simultaneous colloidal processing, and added electrical field from outside, has become interesting. The electrophoretic deposition (EPD) process is one of such ceramic powder assembling technologies. The EPD process is a coating technique using a simultaneous colloidal process and electrochemical driving force[4]. This process is very simple; the dispersed and charged particles in a solvent migrate to an electrode substrate under some potential gradient with a DC power source. The relation of the electrophoretic mobility, v/E (m^2/s·V), is

$$\frac{v}{E} = \frac{\varepsilon_r \, \varepsilon_0 \, \xi}{\eta}$$

(1)

where v (m/s) is the migration speed, E (V/m) is the applied electrical field, ε_r (dimensionless) is the relative dielectric constant of the solution, ε_0 (F/m) is the dielectric constant in a vacuum, η is the viscosity of the solution, and ζ (V) is the z potential of the oxide particles. The EPD technique has some advantages compared with the other fabrication techniques especially for decreasing the fabrication costs and structural flexibility of the substrates[5]. Therefore, the EPD technique has been gaining increasing interest as a ceramic processing technique for a variety of technical applications, e.g., oxide superconductor[6-8], water adsorption/desorption materials[9,10], solid oxide fuel cells (SOFCs) [5,11], insulated coating[12], composite[13], graded composites[14], hydroxyapatite[15], photochatalyst[16,17], optical, catalytic[18], and electrochemical applications[1]. Moreover, experimental results using the EPD technique with several kinds of zeolites (e.g., FAU-, ZSM-5 and so on) as a seeding technique have been reported[19-23].

The solvent of high relative dielectric constant is suitable for the EPD bath because of the principle of electrophoresis. Relative dielectric constant of water is high. Therefore, most EPD researchers that use water based suspension are reported[13,14,17,19,21,22]. In this case, if a high voltage is applied to water, the gas by electrolysis is generated. Therefore, the applied voltage is often low. On the other hand, a high voltage can be applied also to the organic solvent such as alcohol and ketones. A lot of EPD researches that use organic solvent based suspension are also reported [5-12,15,16,18,23]. However, it is necessary to pay attention on handling because these are flammable organic solvents.

In this study, we investigated the technological feasibility in preparation of various ceramic coatings by the applied high voltage EPD in a non-flammable organic bath. Hydrofluoroether (HFE) was applied to the solvent of EPD bath.

EXPERIMENTAL

As the solvent of the EPD bath, HFE from Sumitomo 3M Ltd. under the product name of Novec (registered trademark)[24] series was used. To be more concrete, Novec HFE-7100 (chemical formula: $C_4F_9OCH_3$, boiling point of 61°C, relative dielectric constant of 7.52 and viscosity of 5.8×10^{-4} Pa·s), Novec HFE-7200 (chemical formula: $C_4F_9OC_2H_5$, boiling point of 76°C, relative dielectric constant of 7.35 and viscosity of 5.7×10^{-4} Pa·s) and Novec HFE-7300 (chemical formula: $C_6F_{13}OCH_3$, boiling point of 98°C, relative dielectric constant of 6.14 and viscosity of 1.2×10^{-3} Pa·s) were used as HFE. As the oxide powder, silica (HIPRESICA N3N, Ube-Nitto Kasei Co.,Ltd.), Silicalite[23], Zeolite A4(Tosoh Corporation), Mesoporous silica (Taiyo Kagaku Co., TMPS-1.5) and 8 mol % Y_2O_3 stabilized ZrO_2 (YSZ; Tosoh Corporation, TZ-8Y) were used. EPD bath was prepared by adding the powder to organic solvents, at a concentration of 3-10 g/L. Then, the oxide powder was dispersed by ultrasonic vibration for 10 min. The EPD for fabricating the oxide powder coating was performed using the cell configuration shown in Fig.1. An aluminum plate (length: 50 mm, width: 5 mm, thickness: 0.5 mm) was used as the deposition electrode and two stainless steel plates (SUS304) of the same size were used as the counter electrodes. The electrode distance was 5 mm. For the EPD process, a DC voltage of 100 V was applied for 120 s using a DC voltage current source/monitor (Advantec, R6243). The current density was monitored in situ.

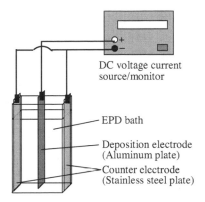

Figure 1. Experimental set up for EPD

RESULTS

As the initial investigation, the HFE-7200 suspensions were used as the EPD bath. The deposition amount on an aluminium substrate was measured as shown in Fig. 2. The amount of deposit was calculated from the weight difference of aluminium substrate before EPD and after EPD (dried). In case of using YSZ, MPS, zeolite A4 and silicalite, the amount of deposit approximately 5 to 10 g/m². In case of using silica, the amount of deposit was approximately 17 g/m². In any case, it was observed that the powder was obviously deposited. Here, silica, silicalite and MPS are almost the same compositions. Moreover, the zeolite A4 is composed of silica and alumina. Therefore, it is thought that these EPD behaviours were similar. However, the specific gravity of silicalite, MPS

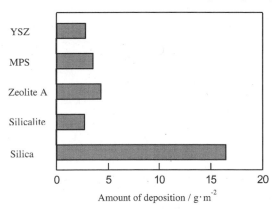

Figure 2. Amount of deposition of each powder when HFE-7200 is used for the EPD bath and applied voltage is at 100V for 120 s

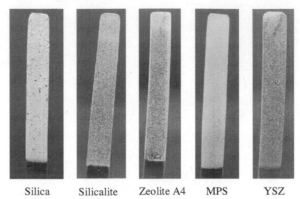

Silica   Silicalite   Zeolite A4   MPS   YSZ

Figure 3. Photographs of the electrophoretic deposited samples of silica, silicalite, zeolite A4, MPS and YSZ by using HFE-7200 and applied voltage at 100V for 120 s.

and zeolite A4 is small because of their porous structure. On the other hand, the specific gravity of silica is large because of the dense structure. Therefore, it is considered that the amount of deposition of silica was large. Fig. 3 shows the photographs of deposited samples. The MPS powder coating was observed completely white. It is considered that the MPS powder was dispersed well in HFE-7200 and it was uniformly coated to the substrate. It is considered that the silica powder also shows similar phenomenon. On the other hand, in case of using silicalite, zeolite A4 and YSZ powder, it seemed that the substrate was observed through the EPD coating. It is considered that the dispersed powder in HFE-7200 aggregated and the aggregated particles were deposited onto the substrate. Still, the powder coated all over the aspects of the substrate.

Here, the effect of HFE was examined about MPS and YSZ whose composition is different. The examinations that used HFE-7100, HFE-7200 and HFE-7300 for EPD baths were carried out. Fig. 4 shows the current density during EPD of YSZ by using three kinds of HFE. The profiles of this current density were almost similar. The notable result is that each value of the current density was very small. The current density was approximately $5 \times 10^{-6} - 20 \times 10^{-6}$ A/m^2. To compare with it, the result of the current density when the 1-propanol was used for EPD of YSZ was shown in Fig. 5. Applied voltage was 100 V, the same as the above experiment. In this case, the current density was approximately 2.0 A/m^2. This value was about six digits larger than that of HFE. Similarly, Fig. 6 shows the current density on EPD of MPS by using three kinds of HFE. The profiles of this current density were almost similar. The current density was also small as well as the case of YSZ as shown in Fig. 4. The current density was approximately $5 \times 10^{-6} - 15 \times 10^{-6}$ A/m^2. To compare with it, the result of the current density when the acetone was used for EPD of MPS at 100 V was shown in Fig. 7. In this case, the current density was approximately 0.1 A/m^2. This value was about four digits larger than that of HFE. It is specified that the current density is small when HFE is used for the EPD bath.

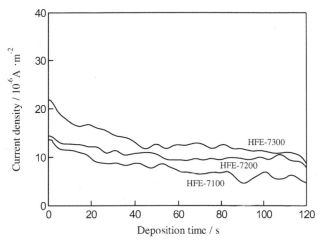

Figure 4. Relationship between current density and deposition time of YSZ powder at 100 V by using HFE-7100, HFE-7200 and HFE-7300 EPD baths

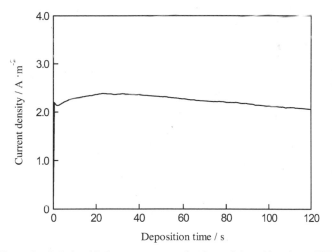

Figure 5. Relationship between current density and deposition time of YSZ powder at 100 V by using 1-Propanol EPD bath

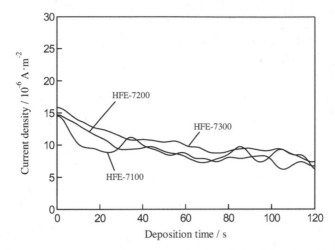

Figure 6.   Relationship between current density and deposition time of MPS powder at 100 V by using HFE-7100, HFE-7200 and HFE-7300 EPD baths

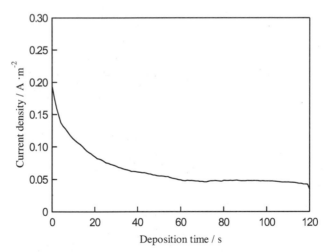

Figure 7.   Relationship between current density and deposition time of MPS powder at 100 V by using acetone EPD bath

The current densities of various EPD were shown in Table1. Mean values of the current density of EPD at initial 120 s were compared under same EPD conditions. In case of using iodine added ketone, such as iodine added acetone that prepared $YBa_2Cu_3O_{7-\delta}$(YBCO) coating[6] and iodine added cyclohexanone that prepared $Tl_2Ba_{1.6}Sr_{0.4}Ca_2Cu_3O_{10+x}$(TBCCO) coating by EPD[8], the current densities are 1.82×10^1 A/m^2 and 9.02×10^{-1} A/m^2 (30 s), respectively. In case of using 1-propanol, that prepared YSZ coating described above, the average current density was 2.23 A/m^2. In case of using acetone, that prepared MPS coating described above, the average current density was 6.44×10^{-2} A/m^2. When HFE was used for the EPD bath compared with these non-HFE EPD bath, current density was smaller from 4 to 7 digits. The electric power that is required for EPD is (applied voltage) x (current density). It was found that using HFE for EPD bath does not only avoid the danger of the ignition of EPD bath but also makes it possible to prepare the oxide powder coating by drastically small electric power.

Table I. Current density of EPD at 100 V

Powder	Solvent	Addition	Average current density of initial 120 s (A/m^2)	Reference
Silica	HFE-7200	none	1.45×10^{-5}	-
Silicalite	HFE-7200	none	1.82×10^{-5}	-
Zeolite A4	HFE-7200	none	9.62×10^{-6}	-
MPS	HFE-7200	none	1.05×10^{-5}	-
YSZ	HFE-7200	none	1.06×10^{-5}	-
MPS	Acetone	none	6.44×10^{-2}	-
YSZ	n-Propanol	none	2.23×10^{0}	-
YBCO	Acetone	Iodine	1.82×10^1	6
TBCCO	Cyclohexanone	Iodine	9.02×10^{-1} (30 s)	8

CONCLUSION

We investigated the technological feasibility of the preparation of various ceramic coatings by EPD in a non-flammable organic solvent. It was found that non-flammable hydrofluoroether (HFE) was useful for the EPD bath for the preparation of various oxide coatings. Ceramic powders, such as silica, silicalite, zeolite A4, mesoporous silica (MPS), and yttria-stabilized zirconia (YSZ) could be deposited on the metal substrate by using the HFE as the EPD bath. In addition, the use of HFE can reduce the power compared to the use of organic solvents such as alcohol and acetone.

ACKNOWLEDGEMENT

This study was supported by Industrial Technology Research Grant Program in 2006 from New Energy and Industrial Technology Development Organization (NEDO) of Japan.

REFERENCES

[1]A. R. Boccaccini and I. Zhitomirsky, Application of Electrophoretic and Electrolytic Deposition Technique in Ceramics Processing, *Curr. Opin. Solid State Mater. Sci.*, **6**, 251–60 (2002).

[2]T. Uchikoshi, T. S. Suzuki, H. Okuyama, Y. Sakka, and P. S. Nicholson, Electrophoretic Deposition of Alumina Suspension in a Strong Magnetic Field, *J. Eur.Ceram. Soc.*, **24**, 225-229 (2004).

[3]E. Dokou, M. A. Barteau, N. J. Wagner, and A. M. Lenhoff, Effect of Gravity on Colloidal Deposition Studied by Atomic Force Microscopy, *J. Colloid Interface Sci.*, **240**, 9-16 (2001).

[4]P. S. Nicholson, in Electrically Active Ceramic Interfaces, Proceedings of US-JAPAN Workshop, pp. 7-14, MIT, Cambridge (1998).

[5]H. Negishi, N. Sakai, K. Yamaji, T. Horita, and H. Yokokawa, Application of the Electrophoretic Deposition Technique to Solid Oxide Fuel Cells, *J. Electrochem. Soc.*, **147**, 1682–7 (2000).

[6]N. Koura, T. Tsukamoto, H. Shoji and T. Hotta, Preparation of Various Oxide Films by an Electrophoretic Deposition Method: A study of the Mechanism, *Jpn. J. Appl. Phys.*, **34**, 1643-1647 (1995)

[7]A. Dassharma, A. Sen and HS. Maiti, Effectiveness of Various Suspension Media for Electrophoretic Deposition of YBCO Superconductor Powder, *Ceramics International*, **19**, 65-70 (1993).

[8]H. Negishi, N. Koura and Y. Idemoto, Electrophoretic Deposition and the Deposition Mechanism of Tl-2223 Superconducting Powder, *J. Ceram. Soc. Japan*, **105**, 351-355 (1997).

[9]H. Negishi, A. Endo, T. Ohmori and K. Sakaki, Fabrication of mesoporous silica coating by electrophoretic deposition, *Industrial & Engineering Chemistry Research*, **47**, 7236-7241 (2008).

[10]A. Miyamoto, H. Negishi, A. Endo, B. Lu, K. Sakaki, T. Ohmori, H. Yanagishita and K. Watanabe, Electrophoretic Deposition Mechanism of Mesoporous Silica Powder in Acetone, *Key Engineering Materials*, **412**, 131-136 (2009).

[11]H. Negishi, K. Yamaji, N. Sakai, T. Horita, H. Yanagishita and H. Yokokawa, Electrophoretic Deposition of YSZ powders for Solid Oxide Fuel Cells, *J. Mater. Sci.*, **39**, 833-838 (2004).

[12]K. Miyazaki, K. Shima, T. Aoki and K. Kamiya, Al_2O_3 coating on Pt wire by the electrophoretic deposition, *J. Ceram. Soc. Japan*, **106**, 1129-1134 (1998).

[13]Z. Wang, J. Shemilt and P. Xiao, Novel fabrication technique for the production of ceramic/ceramic and metal/ceramic composite coatings, *Scripta Materialia*, **42**, 653-659 (2000).

[14]B. Ferrari, AJ. Sanchez-Herencia and R. Moreno, Nickel-alumina graded coatings obtained by dipping and EPD on nickel substrates, *J. Eur. Ceram. Soc.*, **26**, 2205-2212 (2006).

[15]TM. Sridhar, UK. Mudali and M. Subbaiyan, Preparation and characterisation of electrophoretically deposited hydroxyapatite coatings on type 316L stainless steel, *Corrosion Science*, **45**, 237-252 (2003).

[16]S. Yanagida, A. Nakajima, Y. Kameshima, N. Yoshida, T. Watanabe and K. Okada, Preparation of a crack-free rough titania coating on stainless steel mesh by electrophoretic deposition, *Materials Research Bulletin*, **40**, 1335-1344 (2005).

[17]A. Fernandez, G. Lassaletta, VM. Jimenez, A. Justo, AR. GonzalezElipe, JM. Herrmann, H. Tahiri and Y AitIchou, Preparation and characterization of TiO_2 photocatalysts supported on various rigid supports (glass, quartz and stainless steel). Comparative studies of photocatalytic activity in water purification, *Applied Catalysis B-Environmental*, **7**, 49-63 (1995).

[18]KS. Yang, ZD. Jiang and JS. Chung, Electrophoretically Al-coated wire mesh and its application for catalytic oxidation of 1,2-dichlorobenzene, *Surface & Coatings Technology*, **168**, 103-110 (2003).

[19]T. Seike, M. Matsuda, and M. Miyake, Preparation of FAU type zeolite membranes by electrophoretic deposition and their separation properties, *J. Mater. Chem.*, **12**, 366-368 (2002).

[20]B. Oonkhanond and M. E. Mullins, The preparation and analysis of zeolite ZSM-5 membranes on porous alumina supports, *J. Membr. Sci.*, **194**, 3-13 (2001).

[21]L. C. Boudreau, J. A. Kuck, and M. Tsapatsis, Deposition of oriented zeolite A films: in situ and secondary growth, *J. Membr. Sci.*, **152**, 41-59 (1999).

[22]B. Yu and S. B. Khoo, Controllable zeolite films on electrodes - comparing dc voltage electrophoretic deposition and a novel pulsed voltage method, *Electrochem. Commun.*, **4**, 737-742 (2002).

[23]H. Negishi, M. Okamoto, T. Imura, D. Kitamoto, T. Ikegami, Y. Idemoto, N. Koura, T. Sano and H. Yanagishita, Preparation of Tubular Silicalite Membrane by Hydrothermal Synthesis with Electrophoretic Deposition as a Seeding Technique, *J. Am. Ceram. Soc.*, **89**, 124-130 (2006).

[24]http://www.mmm.co.jp/emsd/product/pdt_01_01.html

USE OF GLASS-CERAMIC COATINGS CONTAINING WATER-REACTIVE COMPONENTS AS A BONDING LAYER BETWEEN CONCRETE AND METAL

Weiss, Charles A., Jr., Morefield, Sean W., Malone, Philip G.
USAE Engineer Research and Development Center
Vicksburg, MS, USA

Henry, Karen S., Harte, Sean P.
USAF Academy, CO, USA

Koenigstein, Michael L.
Pro Perma Engineered Coatings, Rolla, MO, USA

ABSTRACT

Vitreous enamels that were designed to bond tightly to metals can be combined with high-melting-point water-reactive silicate- and aluminate-based ceramics to form a bonding layer between a reinforcing metal and an inorganic cement, such as portland or calcium aluminate cement. Vitreous enamels can bond to a variety of metal surfaces, and a variety of water-reactive ceramics also bond tightly to vitreous enamel. The bond strength from steel to enamel is on the order of 35 MPa. The tensile strength of concrete is typically 10% of the compressive strength. The steel-to-enamel bond is over eight times the tensile strength of structural concrete (compressive strength = 40 MPa). The elevated water content of concrete at the metal surface lowers the bond strength into the range of 2.2 to 2.7 MPa thereby making the interface the weakest bond. Pull-out tests were used to demonstrate that using bonding enamel could produce a 4-fold increase in bond strength and make the interfacial bond strength exceed the tensile strength of the concrete. Using a bonding enamel allows the tensile strength at the interface to match the ultimate tensile strength of the concrete. Tests conducted with composite beams demonstrated an increase in strength in experimental beams made with coated versus uncoated steel tubes was observed after 7 days of curing. In sample beams cured for 28 days, the ultimate loading on the coated beams was increased by over 30% on average. The bonding enamel interfaces allows us to take advantage of the new generation of very-high strength concretes in new concrete-steel composites.

INTRODUCTION

The use of reinforcing steel in concrete poses a significant problem in that concrete generally does not bond well to the surface of steel. In typical rebar, the surface of the steel is deformed, and the bar is held to the concrete by the mechanical interlocking of the irregular surface and the surrounding concrete. In straight pull-out tests using rebar with a surface deformation pattern, the concrete is wedged apart by the surface ridges on the surface of the rod, and the concrete cracks[1]. This type of bonding due to mechanical interlocking is less useful in maintaining the bond between steel fibers or sheet steel (steel decking or concrete-filled steel tubes) and concrete[2].

Steel fibers are a good example of the poor performance of smooth steel reinforcement. Steel fibers typically have much smaller aspect ratios than other reinforcement because it is very difficult to uniformly mix long steel fibers in concrete, and it has been shown that hooked fibers will bend to follow the cavity of the concrete as they are pulled out[3]. The problem of making concrete and steel behave as a composite (transferring stress from the matrix component, concrete, to the reinforcement) is also important in the failure of steel decks and concrete-filled steel tubing where the concrete can slide across the surface of the steel as stress is applied.

Nature of the Concrete-to-Steel Interface

Two related problems arise in any discussion of the surface between concrete and steel. Concrete includes water that is needed for hydration of the portland cement and additional water that is needed to change the viscosity of the concrete and make it flow so that it can be placed in forms. As the concrete begins to cure, water not involved in the chemical hydration reactions will separate and move to a free surface. The surface of the reinforcement like the surface of the forms is often where the water not involved in cement hydration coalesces. This bleed water is saturated with calcium hydroxide that often precipitates at the interface. The interface (also referred to as the interfacial transition zone or ITZ) contains very low-density calcium silicate hydrate and abundant platy, crystalline calcium hydroxide. The ITZ is the weakest part of the concrete mass[4].

One consequence of the cement hydration process is that the mass of concrete undergoes shrinkage, whereby the mass of concrete pulls away from the ITZ and the steel reinforcement. As such, the actual bond between the concrete and steel decreases as the concrete ages and cures[5].

Reinforced concrete is typically considered a composite of concrete and steel; but, given a strict definition of a composite indicates that it should be a multiphase material formed from a combination of materials that differ in composition or form, remain bonded together, and retain their identities and properties. The weak bond at the interface of the reinforcing steel can prevent reinforced concrete from behaving as a true composite.

Past Work on the Use of Bonding Enamel

The feasibility of using a reactive vitreous bonding enamel for reinforcing steel in concrete was demonstrated on a laboratory bench scale by the U.S. Army Engineer Research and Development[6]. The results of early rod pull-out tests using smooth steel rods that were glazed with conventional alkali-resistant glass enamel and with the same enamel but with a coating of unhydrated portland cement embedded in the outer surface of the glass enamel showed that bond strength could be increased by using the reactive coating. The hydration of the bonded portland cement layer that occurred when the rods coated with the glass and Portland cement were embedded in fresh mortar increased the bond strength by a factor of 3 or 4 over bare test rods or test rods coated with only glass enamel[7].

Porcelain enamel is one of the most durable coatings that can be deposited on steel. Results of corrosion studies conducted on steel test rods have demonstrated that the enamel coating is effective in preventing corrosion even in damp, brine-saturated sand. In addition, the layer of embedded portland cement improves the inhibition of corrosion. As this embedded cement on the surface of the enamel hydrates, the chemical reaction produces calcium hydroxide, thereby raising the pH of the surface of the concrete. In the event that any imperfections or cracks occur in the porcelain coating, the elevated pH is able to provide passivation protection, and at the same time the calcium silicate hydrate gel from the hydration reaction can fill in defects in the glass[8].

Investigations with mortar-filled steel tubes were undertaken to demonstrate that a new and improved composite could be made using bonding enamel-coated steel. Infilled steel tubes represent an unique opportunity to make a stronger, more durable structural element with a simple coating technique.

FABRICATION OF COMPOSITE BEAMS

Uncoated and bonding enamel-coated composite beams were developed for this testing program. The shells for all of the beams are made by bending 1.21-mm thick (18-gauge), low-carbon sheet steel. Each beam has an octagonal cross-section with an 8.9-cm width and a 6.35-cm height (Fig. 1). Each beam was 81 cm in length and was fabricated by spot welding two channel-like segments at each end.

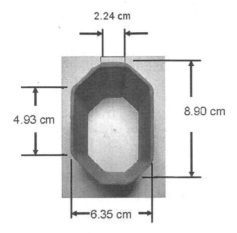

Figure 1. Photo of a coated steel tube or shell formed
from welding two channel-like segments. Each steel
tube form was 81 cm in length.

The coated steel tubes were produced using a single-frit application, single-firing coating system. In preparation of coating, the steel tubes were degreased with a detergent wash, pickled in a mild acid bath, and then treated in a nickel-sulfate bath to condition the surface. The frit as applied as a water-based slurry (slip) that contained a high-zirconium, alkali-resistant, enameling glass, clays, and portland cement (Type I-II). Surfactants in the slip maintained the suspension and retarded the hydration reactions in the Portland cement component. The coated tubes were dried and inspected to make sure that the coating was continuous over both the inside and outside surfaces. The tubes were hung vertically and run through an enameling furnace. The firing times were 2 to 10 minutes, and the firing temperatures ranged from 745 to 850 °C. The uncoated steel tubes were maintained in the condition received after fabrication. No attempt was made to treat the surfaces of the uncoated tubes.

All of the tubes that formed composite beams were fabricated by filling the inside of the tubes with a shrinkage-compensating Portland cement-based grout (Quikcrete Non-shrink Precision Grout, Product No. 158-00). The grout was prepared with a water-to-solids ratio of 0.5. This proportioning produces a one-day unconfined compressive strength of approximately 45 MPa as measured by the ASTM C 109 cube test[9]. The tubes were filled vertically and allowed to cure at 20 °C.

LOAD TESTING OF COMPOSITE BEAMS
Two independent load testing programs were undertaken. The first (preliminary) program was conducted with empty tubes and mortar-filled tubes that were cured for seven days. The second test program used coated and uncoated composite beams that were cured for 28 days.

Seven-day Test Program
The samples that were cured for 7 days were tested to determine the flexural strength using a protocol similar to that outlined in ASTM C 78-02, Standard Test Method for Flexural Strength of Concrete (Using a Simple Beam with Third Point Loading)[10]. The test set-up used a support span of 22.9 cm and a nose span of 7.6 cm. Flexural strengths were calculated using the Instron-brand, Partner software.

Twenty eight-day Test Program

The samples cured for 28 days were also tested according to the ASTM Standard Test Method C78. The test set-up used an 81-cm support span and a nose span of 26 cm. The tests were conducted with a MTS Testing Machine, and the test results were presented as the ultimate load for the beam.

RESULTS AND DISCUSSION
Seven-day Program

The results from the flexural strength tests from the 7-day-cure program are presented in Table I. For each type of sample, the table contains the lowest and highest values of flexural strength obtained from the tests, the mean value from at least three identical test specimens, and the standard deviation about the mean.

Table I. Comparison of Flexural Strengths for Coated and Uncoated Empty Tubes and Concrete-filled Composite Columns

Sample Type	Lowest Flexural Strength (MPa)	Highest Flexural Strength (MPa)	Mean Flexural Strength (MPa)	Std. Deviation (MPa)
Empty tube uncoated	2.93	3.94	3.60	0.58
Empty tube coated	3.57	5.40	4.47	0.92
Concrete-filled, uncoated tube	20.4	37.3	31.5	9.60
Concrete-filled, coated tube	30.3	43.9	35.0	7.75
Increase in Avg. Flexural Strength for Coating (Empty tubes) = 0.87 MPa				
Percentage Increase in Avg. Flexural Strength for Coating (Empty tubes) = 24				
Increase in Avg. Flexural Strength for Coating (Filled tubes) = 3.47				
Percentage Increase in Avg. Flexural Strength for Coating (Filled tubes) = 11				

The data from the empty tubes indicate that the enamel coating of the steel increases the flexural strength. Improvements in the strength of steel have been documented when vitreous enamel is applied over the steel[11].

Measuring flexural strength of a composite is complex because the bending of the steel accounts for part of the strength, and the strength of the concrete core alone introduces another variable. As with composites in general, the strength of the bond that allows the stress to be transferred to the steel is an additional factor. Once the concrete core breaks, the beam can only yield by stretching the steel skin on the bottom of the beam or by extruding the concrete core from the interior of the composite or both.

Twenty-eight Day Program

The results from the peak loading tests from the 28-day-cure program are presented in Table II. For each type of sample, this table also contains the lowest and highest values of flexural strength obtained from the tests, the mean value from at least three identical test specimens, and the standard deviation about the mean.

Table II. Comparison of Peak Loadings for Uncoated and Coated Composite Beams

	Lowest Peak Loading (kN)	Highest Peak Loading (kN)	Mean Peak Loading (kN)	Std. Deviation (kN)
Uncoated steel	12.13	15.19	13.97	1.62
Coated steel	16.04	21.83	18.46	2.99
Increase in Average Peak Load for Coating = 4.49 kN				
Percentage Increase in Average Peak Load for Coating = 32.1				

In the case of the composite beams that cured for 28 days, the difference in the ultimate yield (32% increase for the coated beam) indicates that the difference in bond strength is an important factor. An examination of the ends of the beams indicated that the concrete cores were extruded when the composite beam yielded (Fig. 2a and b).

a)   b)

Figure 2 a. End of the composite beam made using an uncoated steel tube. The failure shows the smooth concrete-steel interface. b. End of a composite beam made with a coated steel after loading. Note that when the core was extruded, it sheared off the black enamel coating on the interior of the tube indicating the failure probably occurred in the enamel coating.

CONCLUSIONS
The results from these initial tests with composite beams made with coated and uncoated steel tubes indicate the following.

1) An increase in the strength of a steel-concrete composite beam can be produced by using a coating that consists of a vitreous bonding enamel containing Portland cement.
2) The effects of the bonding coating on flexural strength can be seen after as little as seven days curing.
3) It is possible to significantly alter the concrete-steel interface by adding the vitreous bonding enamel, and after a 28-day of cure time, up to a 30% increase in ultimate flexural load can be produced.
4) Failure of the bonding coating appears to occur in the enamel and not at the concrete-enamel interface.

ACKNOWLEDGEMENTS
The funding for this study was provided by Office of the Under Secretary of Defense (Mr. Dan Dunmire and Dr. Lewis Sloter) and ACSIM-IMA (Mr. David Purcell and Mr. Paul Volkman).

REFERENCES
[1]E. G. Nawy, Reinforced Concrete, A Fundamental Approach 2nd Ed., Prentice-Hall, New York (1990).
[2]M. Maage, Fibre Bond and Friction in Cement and Concrete, RILEM Symp. On Testing and Test Methods of Fibre Cement Composites. The Construction Press, Hornby, England, Paper 6.1, 329-336 (1978).
[3]M. Maage, Interaction between Steel Fibers and Cement Based Matrixes, *J. Mat. and Struct.*, **10**, 297-301 (1977).
[4]Bentur, S. Diamond, and S. Mindess, The Microstructure of the Steel Fibre-Cement Interface, *J. Mat. Sci.*, **20**, 3610-20 (1985).
[5]Fu and D. D. L. Chung, Decrease of the Bond Strength between Steel Rebar and Concrete with Increasing Curing Age, *Cem. and Conc. Res.*, **28**, 167-9 (1998).
[6]L. Lynch, C. A. Weiss, D. Day, J. Tom, P. Malone, P. Hackler, and M. Koenigstein, Chemical Bonding of Concrete and Steel Reinforcement using a Vitreous Enamel Coupling Layer, Proceedings of the 2nd International Symposium on Connections between Steel and Concrete, 425-38 (2007).
[7]V. Hock, S. Morefield, C. Weiss, D. Day, J. Tom, P. Malone, C. Hackler, and M. Koenigstein, The Use Of Vitreous Enamel Coatings to Improve Bonding and Reduce Corrosion in Concrete Reinforcing Steel, Corrosion 2008, Proceedings of the NACE International Conference, New Orleans, LA, Paper 1264 (2008).
[8]S. W. Morefield, V. F. Hock, C. A. Weiss, Jr., P. G. Malone, and M. L. Koenigstein, Reactive Silicate Coatings for Protecting and Bonding Reinforcing Steel in Cement-Based Composites, Proceedings of the U. S. Army Science Conference (2008).
[9]American Society for Testing and Materials, ASTM C 109C-109M – 07, Standard Test Method for Compressive Strength of Hydraulic Cement Mortars (Using 2-in. or [50-mm] Cube Specimens), ASTM, West Conshohocken PA (2007).
[10]American Society for Testing and Materials, ASTM C 78-02, Standard Method for Flexural Strength of Concrete (Using Simple Beam with Third-Point Loading), ASTM, Conshohocken, PA (2002).
[11]F.Nagley, Engineering properties of porcelain enamel in Conference Proceedings Porcelain Enamel in the Building Industry, November, 1953, National Academy of Sciences—National Research Council, Washington, DC (1954).

Author Index

Author Index

Sjöström, S., 23
Snead, L. L., 147
Sun, J. G., 87

Tan, Y., 75

Ueta, S., 147

Vasudevamurthy, G., 147

Wang, X. W., 123

Watanabe, K., 177
Wei, C., 95
Weiss, C. A., Jr., 187
Wynick, G., 123

Yanagishita, H., 177
Yilmaz, S., 135
Yongjie, W., 95
Yücel, O., 135

Zhu, D., 75